SCIENTIFIC REASONING
AND EPISTEMIC ATTITUDES

SCIENTIFIC REASONING AND EPISTEMIC ATTITUDES

BY

LÁSZLÓ HÁRSING

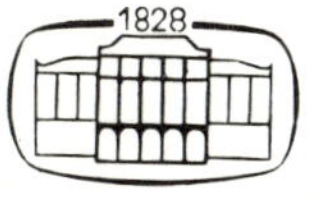

AKADÉMIAI KIADÓ, BUDAPEST 1982

Translated by

BALÁZS DAJKA

ISBN 963 05 3003 1

Printed in Hungary
Akadémiai Nyomda, Budapest

CONTENTS

PREFACE

This brief monograph uses elementary logic and mathematics as its means to reconstruct the structure of rational scientific reasoning. It analyzes the presuppositions which underlie the rationality of reasoning and the factors which determine the field of reasoning as a homogeneous medium.

Among the factors examined, distinctive roles are assigned to the logical probabilities of hypotheses (theses of argumentation) and to arguers' epistemic attitudes which reflect the basic epistemological features of the reasoning situation. Expected epistemic utility functions are determined by epistemic attitudes and by logical probability as the degree of the relative truth (plausibility) of hypotheses. These functions allow us to elaborate reasoning strategies more adequate than either plausible inferences or rules of acceptance based on the degree of confirmation, systematic and explanatory power, or the degrees of corroboration and cogency.

Several other non-classical types of logic are also suitable for handling plausible inferences, but their adequacy falls far behind that of probabilistic logic.

I express my gratitude to Imre Ruzsa and Sándor Szalai for their valuable help and advice, and to Márta Fehér for her interest in my work.

László Hársing

1 INTRODUCTION

The objective of my work is to examine the logical nature of scientific reasoning. My method may generally be described as *applied logical research*, my cognitive goal having essentially been to scrutinize the rules of scientific reasoning in the light of recent results in symbolic logic.

Applied logical research is especially important because, as it were, it feeds back findings of theoretical research into the actual process of cognition. It provides new elements of knowledge which not only lay claims to abstract truth but may also bring about an increase in the effectivity of cognitive processes. Applied logical research resembles applied mathematics to a considerable extent, even though it differs in its wider range, as it is successfully utilized in exploring problems which are as yet extremely difficult to formulate in a strictly quantitative way.

At the same time, this approach may also be taken as pertaining to the *philosophy of science*, for it is directed at the logical reconstruction of a certain process of scientific cognition. Owing to the very nature of the problem under investigation, this mode of analysis is penetrated and deeply influenced by epistemological considerations.

In general, the phenomena under investigation cannot directly be subjected to any analysis. Instead, the original objects have to be replaced by their more or less idealized models, a method we must also follow throughout our logical analysis of scientific argumentation.

An act of scientific reasoning is always part and parcel of the science of the age, influenced by a host of historical, psychological, sociological, communicative, etc. phenomena. Reasoning conceived as concrete and distinct will be termed *actual reasoning*. Actual reasoning is a special form of communication designed to make others share one's own beliefs. Parents, teachers, journalists, politicians or scientists would scarcely be able to perform their social roles without possessing suitable capacities of reasoning. When performing one of these roles, in accordance with the actual situation and the object of argumentation as well as the personal traits of its participants, an individual may resort to the most varied reasoning procedures which all can be classified under the notion of actual reasoning. We have no space here to dwell on the characteristics of this notion in detail. (Cf. Konger, 1960.) Nor can we undertake supplying even a sketchy analysis of the main types of some more important patterns of reasoning found in the field of scientific communication. It seems to be sufficient to draft a phenomenal description of *scientific discussion* as one of the most frequent and best known communicative acts, drawing upon a stimulating study by Ferenc Pataki (Pataki, 1977, 338—353).

Whoever undertakes a creative contribution to science will scarcely be in the position to refrain from scientific discussion. Not only is his debating ability most telling of the level of intelligence and professional competence of a scientist, but he is also urged to participate in discussions first of all by his wish to become thoroughly familiar with, or convince others of, newly emerging ideas. Concomitant intellectual excitement and the pleasure of challenging rivals are no less sure to play a part in kindling scientific controversy.

Scientific discussions are widely and justifiably called a 'functioning scientific public opinion', a scene of testing alternative responses to scientific problems. In their absence, the scientific public opinion would certainly lose its orienting and evaluating functions.

The publicity of discussions, and the freedom to maintain one's position are basic criteria of democracy in scientific life, even if they are frequently counteracted by the particular interests of certain research groups. There are cases when ideological or misconceived political considerations interfere and lead scientific debates in the wrong direction.

As it is well known, a significant portion of discussions in science cannot be credited with any results, although the broader public reasonably expect scientists to set a good example in the fruitful interchange of ideas. In relation to our subject, it is perhaps not out of

10

place here to take a glimpse at some of the main factors that exert a negative effect on the success of scientific debates.

It is not unusual in science for debates to become a *substitute for research*. In these cases, disputes function as a surrogate and, most probably, they are doomed to failure from the very outset, for it is a prerequisite of any success in discussions that rival views be sufficiently mature. Properly elaborated conceptions, however, are more likely to be an outcome of prior research than of subsequent discussion, even though a confrontation of ideas is often a suitable means of more precisely formulating opposed views. The true point of disputes is therefore misunderstood by those who believe that mere participation in them is a scientific action. They seem to forget that discussion is a mere device, which has as its function to provide participants with new and true knowledge. Should disputants fail to bear this in mind, their clash will become sterile polemic capable only of concealing the lack of mature and substantial thoughts. Most disputants feel a deep-seated conviction that *the position they have taken is absolutely right, whereas their opponents are wrong once and for all*. Given these circumstances, it is very difficult to carry out any scientific reasoning when the essential condition of a dialogue is not fulfilled, viz. the condition that requires participants to be able to place themselves in the position of their adversaries, i.e. to base their criticism not only on their own grounds but, by hypothetically taking their opponent's position, to reveal the inherent weaknesses of that position, too. Otherwise the debate is soon reduced to a mere clash of opinions, leaving no chances for reconsideration, self-correction and a dialectical synthesis. It breaks down to mutually exclusive monologues.

Among scientists or scholars, in many cases an exchange of ideas takes place with the *appearance of a debate* when in fact the majority of the participants have previously agreed, at least tacitly, upon the alternative positions each of them will take. A discussion so staged will generate no feeling of interest and the disputants will play their parts as if trying to perform a prefabricated script, while organizers are inclined to interpret the statements and suggestions proposed during the discussion as arguments supporting the previously arranged plan of decisions.

The distortion of scientific debates comes to a considerable extent from *previously developed false attitudes*. These attitudes become fixed as prejudice and prevent the consideration of each other's arguments, often only facilitating the incorporation of new knowledge if it is in accordance with them.

In certain cases, especially in the field of social science, *the real emphasis is not on the actual subject matter of the debate* but on the hidden motives and intentions that underlie it. These discussions can only be followed by the initiated, whose ears are quick enough to notice what the controversy really is about. This type of dispute is usually interwoven with latent hostility which the participants feel but are reluctant to reveal.

Scientific debates as they actually take place are perhaps most often distorted by the fact that *some participants argue for or against persons instead of opinions*, and thus replace the evaluation of scientific achievement with the evaluation of persons. There are reasoners who regard discussions as beneficial to their professional careers and are therefore ready to stand up compliantly for cases contrary to their own conviction. Others may refrain from taking sides because they prefer not to be associated with disputants whom they happen to dislike. In extreme cases of being personal, some participants may be inclined to use debates as a means of eliminating their opponents from scientific life altogether.

This rather sketchy account of actual disputes in science cannot constitute but only outline a description of the clusters of themes and divisions of fields in concrete research that are studied by the science of science, the history of science, the sociology and psychology, the ethics and informatics of science. It also attempts a bird's-eye view of that area of reality of which *rational scientific reasoning* is an *abstract model*, or an *ideal type*.

In order to progress from the sphere of actuality to that of rationality representing the ideal, it is necessary to satisfy certain preconditions. Let us consider in the following those concrete *preconditions of idealization* which are to 'transfer' actual (and, consequently, not always rational) situations of reasoning to the sphere of rationality, facilitating in this way the *homogenization* of scientific reasoning to a degree that it may reveal the logical relationships beyond epistemological ones.

Here the relative beginning of the construction of scientific theories is identified as the creation of a *homogeneous medium* through idealization and abstraction from the heterogeneous, empirically given concrete. As Georg Lukács writes: "Ein homogenes Medium der Wissenschaft kann nur aus einer bereits — relativ — erfaßten Realität selbst gewonnen werden. Seine Grundlage bilden Elemente und Zusammenhänge der objektiven Wirklichkeit selbst, deren abstrahierende Bearbeitung, eben das Schaffen eines solchen homogenen Mediums, vor allem darin besteht, das objektiv Seiende von jeder an die Subjektivität gebundenen, anthropomorphisierende Tendenzen enthaltenden, Betrachtungsweise nach Möglichkeit zu reinigen" (Lukács, 1963, vol. 1, 642). However, the abstract apprehension of an object in the first step of theory construction already implies the eventual accomplishment of the possibility of returning, on a higher level, to the concrete, heterogeneous object. The history of the development of science demonstrates how cognitive human mind proceeds from the empirically given, more or less crudely concrete object of cognition to its abstract apprehension and then, using these abstractions as elements in thinking, to reconstructing the empirically given concrete. By rising, so to speak, above reality, the human mind becomes capable of forming a picture of this reality as a whole from a detached perspective, apprehending at the same time the deeply rooted, essential relations of the object.

We apply several levels of abstraction to elaborate the reality directly given in experience and to prepare it for theoretical consideration. In *ordinary abstraction*, figuratively speaking, we strip the object of its prima facie accidental properties. The resulting concepts, however, prove to be too crude, vague, and complicated for theory construction. Therefore we reconstruct them in thought. A concept resulting from ordinary abstraction may entail some characteristics which we deem especially important from a theoretical point of view besides others which seem rather to impede than to advance the construction of theories. By means of the second level of abstraction, *idealization*, the first kind of characteristics will in a way be "intensified" and the second "suppressed" in our thoughts. Finally, there is a third step of idealization which we call the *ideal restriction* of change, in which we assume certain variable characteristics to be constant or varying only within a definite interval.

The same steps of idealization are taken in creating the concept of "rational scientific reasoning". To begin with, we single out and identify scientific reasoning as a phase of cognition. Such ordinary abstraction leads us to the concept of "actual scientific reasoning", still intertwined with historical, sociological, and psychological factors. This level of abstraction makes it possible to work out a phenomenology of scientific reasoning, with extensive case studies into the historical, sociological, psychological, etc., aspects that appear significant. We may seek answers to questions like "What are the empirical properties of scientific reasoning as a specifically refined, theoretical type of social practice?" We can obtain qualitative and quantitative generalizations. The more profound relations inherent in our object of study, however, cannot be revealed unless abstraction is further extended to arrive at a model, now on the ideal level, of scientific reasoning, which we term "rational scientific reasoning".

What has hitherto been put forward is meant as a theoretical grounding for the preconditions of idealization to be introduced below. We suggest that a study of these preconditions will save us from the dangers of both extreme empiricism and extreme formalism. Concepts obtained through idealization are of a partly empirical, and partly theoretical nature, being models that can be used to substitute a given area of reality, but nevertheless always displaying features which embody some prior expectations of the human mind. Therefore, idealization is a rather efficient process of the logical reconstruction of reality. Without its help the borderline between the philosophy of science and the particular sciences studying science could not be properly drawn.

The task of determining the conditions of the rationality of scientific reasoning does not seem to be an easy undertaking. In this case, we are required to answer the question "Which are the basic characteristics of 'good' or proper scientific reasoning?" An answer like this may be supplied: reasoning is rational if, and only if, reasoners argue like scientists really should, and free themselves from all other considerations alien to the scientific attitude. This is the kind of answer given by Harrah (Harrah, 1963, 9—10). However, this answer is analytical as it implies that we have defined the concept "scientific" with sufficient precision. This assumption is highly conditional, for the concept of what is "scientific" is subject to historical change and its contents are too complex to allow exhaustive analysis. It may prove to be a more fruitful attempt at an answer if we turn to some essential criteria of what is scientific.

a) The rationality of reasoning presupposes a sufficient degree of the *maturity of scientific ideas.* Discussing half-finished, immature views will most probably exert a harmful effect on the eventual progress and future commensurability of these conceptions. If a given phase of cognition requires the predominance of research, any efforts at controversy over different incomplete solutions to a problem are irrational.

b) As various proposed solutions to a problem are confronted in scientific debates, one can hardly accept as an epistemological principle the existence of but one single and indivisible truth. It is therefore prerequisite to the rationality of the debate that the mutually exclusive alternatives of true and false be replaced by a *continuum of relative truth ranging from the totally* (surely) *true to the totally* (surely) *false.* The aim of scientific discussion is not to discover the absolute and final truth but, instead, to select and accept the optimum alternative of truth in the given cognitive situation. A dogmatic approach claiming the ultimate truth for itself will scarcely be one that invites truly scientific discussion.

c) An important requirement for the rationality of reasoning is *full democracy in science.* Accordingly, mere *prestige argumentation* is excluded from scientific reasoning. The fact that a thesis is proposed for discussion by a scientist of distinction and therefore of great authority can in no way exempt the scientific public from its duty to supervise it. This principle leaves no opportunity for situations to arise in which the right to put forward new ideas is reserved for certain scientists while others are limited to participate in the debate over it. A truly democratic scientific public allows no place for scientific dictatorship, however great previous achievements are invoked in attempts to impose it. It is equally impossible for public life to be democratic without wide *publicity*, which means the same as the requirement of *intersubjectivity*. Sufficiently reliable control of the theses under discussion and unhampered communication within the given scientific community can only be guaranteed in this way. Finally, a democratic scientific public life also implies the *principle of mutualism*, which states that no scientific reasoning is rational unless opponents have some opportunity to expose their arguments and appeal to others for acceptance.

d) We suppose, furthermore, that reasoners have become detached from the prejudices and roles imposed by their social status to an extent that enables them to recognize the ultimate aim of knowledge as *the promotion of the intellectual progress of mankind*. Otherwise they cannot even be considered as scientists in the noble sense of the word. These requirements are fulfilled more easily in the natural sciences than in the social sciences, though the latter also display tendencies towards satisfying these criteria.

Activity in the interest of the intellectual progress of mankind is closely tied up with the principle of *objectivity*. This fundamental norm of science prescribes that our knowledge should be the most objective possible, that is, its contents must be basically determined by the object studied and never influenced by particular human interests, while its form should preserve the contents without distortion.

The rationality of scientific reasoning, however, need not exclude the use of presuppositions in research. It only purports to fend off the effect of presuppositions that have become prejudices and therefore false.

e) Reasoning could hardly be thought of as rational without the *creative nature* of debate. Without this, reasoning is just a guise for fake discussion, in which all the participants have given up their efforts from the outset to single out the optimum among the alternative solutions to the problem in question, refraining from discovering the truth attainable at the given level of knowledge, which is, therefore, relative. Only creative debates can transcend the particularity of any given community and rise to the level of general humanity.

f) The rationality of reasoning requires all participants to fully observe *the ethical norms of research and of publishing* its results (Merton, 1972, 65—69). These norms are of course historically relative according to the nature of each science, but, in any case, they are definite enough to circumscribe both the *purity of scientific life* and the *personal integrity of the scientist* (Lakatos, 1971, 92). The first of these postulates, to put it perhaps too simply, is taken to mean that in making their value judgements, the participants should remain perfectly independent of circumstances merely incidental to scientific reasoning so that their opinions are not influenced by anything incompatible with scientific cognition (however important it may be in other respects). The second condition, stipulating the integrity of the scientist, is met if his means of argumentation are fully honest in this specific intellectual combat, the "game" of cognitive strategies.

g) The condition of *minimum agreement* is the last to be mentioned and will find the widest application in what follows later. This requirement will hardly need further justification if we consider how it would be possible to successfully argue if there is not at least partial agreement among reasoners upon the answers to the following questions: What are the theses (hypotheses) under discussion? What are the arguments for and against them? What new arguments may be needed for empirically testing these hypotheses? What logical laws regulate the connection between hypotheses and arguments? What logical values are used to characterize their relations? What are the rules, if any, of valuation? How are value judgements influenced by the specific features of the reasoning situations? Without a degree of agreement upon questions like these, the interlocutors can hardly be thought of as striving to enrich the store of human knowledge when their differences in learning and attitudes make it hopeless for them to define the body of knowledge which is not disputed within the limits of the given discussion or to agree upon permissible ways of reasoning. In order to avoid failure, they must make mutual concessions and attain partial consent as to the judgement of all factors that determine or at least influence the conclusion of scientific reasoning. A rational scientific arguer is characterized by a sincere effort to understand his opponent's concept without giving up his own, and by an ability to think things out to an extent from the point of view of his partner. (See Apostel, 1964, 305—307.)

We do not assert that reasoning is rational whenever the above criteria are fulfilled, but perhaps we may say that, if they are fulfilled, scientific argumentation can be considered rational with a high probability. Throughout the following chapters the rationality of reasoning will always be stipulated and, accordingly, it will be assumed that the criteria listed under *a* to *g* are satisfied. These considerations should keep us clear of confusion when for the sake of simplicity we use the terms "scientific reasoning" or "reasoning".

That the above-mentioned idealization is not an arbitrary act of thinking may be demonstrated by the fact that scientific public life itself displays an analogous process. A new hypothesis is usually discussed for the first time by small scientific communities. Argumentation here may be heavily distorted by factors of no scientific importance. If, however, the arguments and counter-arguments are submitted to a nationwide or an international scientific public, it can not only be expected that the disputed suggestion will undergo selfpurification but, more and more, that non-scientific considerations will neutralize each other through an averaging process. Therefore it may not be an overstatement that the rationalization of scientific reasoning, or at least a tendency, occurs in scientific life itself.

These presuppositions are nevertheless too abstract to represent scientific reasoning with clear outlines as an idealized cognitive situation. We must, if partially, break down the abstract character of our approach and examine what determines rational scientific reasoning.

We have until now delineated the subject matter of our study negatively rather than positively, suggesting presuppositions of idealization to clear it from superficial and irrelevant features. Now we shall subsequently examine the constituents which serve as the elements of argumentation. The ordered set of these elements will be called the *field of reasoning*. The field of reasoning is a thought construction containing all the factors that are relevant for the theoretical study of reasoning as an idealized cognitive situation, and it is also a homogeneous medium in which the valuation of hypotheses takes place in a form that is adequate and liable to theoretical reconstruction. We shall first deal with hypotheses for they constitute the objects proper of reasoning, i.e. the *theses of argumentation*.

2 HYPOTHESES

Our cognitive goal in arguing is to add to human knowledge. The new knowledge which is supposed to enrich previous knowledge will be termed a *hypothesis*. Hypotheses do not as yet constitute real knowledge. They rather represent possible knowledge and, as such, they have a chance to rise to the status of scientific truth. Hypotheses are therefore "candidates for truth".

Regarding hypotheses from the aspect of their origins, they must be approached from the basis of *scientific problems*. These problems demarcate areas where knowledge is lacking, where there is a gap between acquired knowledge and the actual cognitive needs. Hypotheses purport to fill this gap. In this way, they can be considered as *possible answers* to problems that have emerged.

As usual, problems may receive more than one possible answer, but the number of the hypotheses that in fact decrease or eliminate the lack of knowledge manifested by the problems is very often quite limited. Cases considered as typical will be the ones in which two hypotheses are in competition to provide the answer to a certain problem. These are called *rival hypotheses*. Rival hypotheses constitute alternative, incompatible or disjunct answers to some problem. In a way, they delineate the range of possible answers at any actual stage of cognition. Alternative cases permit at least one answer, incompatible cases permit no more than one, while disjunct ones permit precisely one possible answer. The simplest case of rival hypotheses is obtained by contrasting a hypothesis as a possible answer with its negative. We must note, however, that a mere negation is not in fact an answer. It is just an overall allusion to the logically possible alternative answers.

In order to establish a logical theory of scientific reasoning, it is not necessary to go into details of classifying hypotheses. It will suffice to mention the kinds of hypotheses that are the most frequently found in scientific reasoning.

One possible way of enlarging human knowledge is to extend our existing knowledge over all similar cases (generalization) and to try to find necessary connections between events or their determinations by systematically excluding random alternations. (Here, in a somewhat crude formulation, a phenomenon which regularly and therefore certainly recurs is said to be necessary.) This is how we come to set up *nomic hypotheses* supposing that certain necessary relations hold within a definite circle (universe) of phenomena. They provide a way to code previously acquired and isolated knowledge in a generalized and methodically contracted form. Nomic hypotheses usually cover a very great or even an infinite number of states of affairs, scattered in distant zones in space and time of the world. They enable our knowledge to expand and push back the limits of ignorance at any given time. It is thanks to these hypotheses that retrodiction of past events and prediction of future ones take place. As for their form, they are simple statements of a structure approximately represented by the formula "in the universe of discourse U, every phenomenon which has the property A has the property B too".

It is a difficult problem in the philosophy of science how to differentiate between nomic hypotheses and ad hoc generalizations such as "All the apples in this basket are red", or "It was windy every day between 20th and 25th January, 1978, in Budapest" etc. In general, lawlike statements are required to hold for unlimited or infinitely large sets of events. Accordingly, nomological hypotheses are statements made of events occurring in any point of space and time in some space-time interval. We are fully aware of the fact that this brief observation can be no more than a hint at the nature of an answer to our problem, but a more detailed account would carry us too from our subject.

An even more important part than that of nomological hypotheses is played in scientific reasoning by those which claim to be theoretic i.e. *theoretical hypotheses*. They have as their basic purport to organize

20

isolated or loosely connected nomological and other hypotheses into a unified system of thought. They are termed theoretical because they contain concepts like the idealizations examined in the foregoing, as well as thought constructs formulated after objects of reality that are not directly accessible to experience. Such thought constructs are e.g. "radioactivity", "half-period", "negative entropy", "absolute monarchy", etc. These scanty hints are merely meant to indicate the frame of reference in the philosophy of science which provides the conceptual scheme of our problematic. What is most significant from the angle of the theme in question is that all actually known scientific theories started their careers as complex theoretical hypotheses before they were reinforced by arguments and favourable reasoning situations eventually to become theories.

It is one of the most important tasks of both nomological and theoretical hypotheses to furnish *explanations* to a more or less wide range of empirical facts. Nomological hypotheses refer to the general and the necessary in explaining the individual and the accidental. Theoretical hypotheses, on the other hand, undertake the reconstruction in thought of the deep interrelations in the realm of reality under examination in order to explain more or less verified nomological hypotheses (empirical regularities) or statements describing important individual states of affairs. The former case is mostly exemplified by the natural sciences, the latter rather by the social sciences. The usual way of making up explanatory hypotheses in natural science is to reveal the so-called inner structure or mechanism of the sphere of phenomena under consideration, and then to point to it as the reason for the emergence of the explained lawlikeness as a functional determination representing a general mode of behaviour. In the social sciences, causal hypotheses are first of all applied to explain individual facts. It may be observed that nomological hypotheses have the effect of widening our knowledge while explanatory hypotheses mainly contribute to deepening it. A rather simplified way to formulate the structure of explanatory hypotheses may be by the proposition "*B* because *A*", where *B* may be called the *explanandum* and *A* the *explanans*.

For us, however, the most important feature of hypotheses in the given respect is perhaps that they cannot directly be confronted with reality, only through the mediation of further knowledge. Ultimately, here lies the explanation of why it becomes necessary to work out the logical foundations of hypotheses, a process taking place within the framework of reasoning. This problem cannot be properly discussed unless the study of background knowledge is also brought under the scope of the present investigation.

3 BACKGROUND KNOWLEDGE

Scientific reasoning is a specific kind of valuation, hypotheses as theses of argumentation being confronted in it with knowledge the positive epistemic value of which we are convinced. Items of this knowledge will be called *arguments*, their system *background knowledge*.

Arguments provide the logical basis on which to accept some hypothesis, and they cannot function as a firm basis unless their epistemic value is positive beyond doubt. A positive epistemic value derives from correspondence to reality. If our knowledge offers a true picture of a certain sphere of reality, we call it *true*. If the degree of correspondence is insufficient and the picture is distorted, we speak of *falsity*, which is regarded as a negative epistemic value.

"True" and "false" are poles asunder as a pair of values of which as a rule one and only one may hold for any knowledge. Arguments, and their complex called background knowledge must be supposed to be true. In the actual case it is of no importance how precisely they represent reality. We are content to know that on any given level of cognitive development the correspondence is satisfactory, even though it is never perfect. Obviously, it may happen that the development of cognition results in a demand for our knowledge to be more adequate to reality than before, and thus part of the knowledge we used to hold true will have to be deemed false. Therefore, we cannot speak of background knowledge in general but only in relation to some definite stage in the development of science, i.e. considering it to be historically relative. Should some apparently true knowledge later prove false against the standard of science, it may still correspond to reality well enough to be widely applicable in some fields of non-scientific cognition (everyday knowledge, technological knowledge, etc.).

In order to throw more light on the cognitive status of background knowledge, we must require it *not to be logically true:* it should be true on account of its correspondence to reality and not in its mere logical structure. Because the degree of its correspondence to reality is historically relative, background knowledge always involves a measure of falsity as well. Nevertheless it is required *not to be logically false* i.e., self-contradictory.

Background knowledge includes all the *specific experience* relevant to the epistemological judgement of the hypotheses put forward as theses of argumentation. The great bulk of this experience is in a way found ready by successive generations in history, which attain it by *communicative learning* through the mediation of their predecessors.

Learning does not begin with doubting; it proceeds from the more and more complete reproduction of available experience. Otherwise we could not acquire the basic knowledge necessary for us to become cognitive subjects of full capacity. Doubt is always dependent on knowledge, and not vice versa. In effect, this may make it clear that we cannot doubt the truth of the complete prior knowledge, yet we could hardly point to any bit of knowledge that could not be doubted for its adequacy in an actual cognitive situation. The principle "de omnibus est dubitandum" is only valid if supplemented by the clause "nunquam de totalitate scientiae".

At the outset, acquisition is basically a reproductive process which, throughout the shaping of a fully developed cognitive subject, will assume a more and more creative character and will be transformed into an act of task solving, which is in its turn preparatory for problem solving.

Problem solving is ultimately no less than the intellectual appropriation of reality on a high level, an activity often called *learning from nature.* Background knowledge includes among others true knowledge acquired independently by reasoning cognitive subjects so as to attain better grounds for their value judgements concerning the hypotheses in question.

A distinct kind of reasoning situation is one in which one or more hypotheses are related to the same background of knowledge as a logical basis in order to find out the degree to which these hypotheses are corroborated by the background knowledge, or how they could be ranked according to their degree of corroboration. In such cases we speak of *synchronic* reasoning situations. However, it frequently happens that the changes in the degrees of corroboration of hypotheses are related to several backgrounds of knowledge following each other in a time sequence. These situations of reasoning are called *diachronic*. The synchronic approach can be considered a borderline case of the diachronic method of investigation.

In diachronic situations of reasoning we can distinguish between *initial* (prior, a priori) and *later* (posterior, a posteriori) *backgrounds of knowledge*. Initial background knowledge will henceforth be designated with a k. Naturally, in any case we can only speak of a relative beginning, each earlier background qualifying as initial in contrast to a later one. The "differences" between initial and later backgrounds of knowledge will be called *additional knowledge* or supplementary information.

Initially, the additional knowledge p is itself a hypothesis. Whether of nomological or of theoretical character, it lends itself to empirical testing more easily than the hypothesis h. The problem of how the empirical testing of the additional knowledge p may be executed will not be handled here. We merely observe that ultimately the verification or refutation of p is to be effected directly or indirectly through confronting it with empirical data (Bunge, 1967, Vol. 2, 290—347).

If p or $\sim p$ is known to be true, we link it to the initial background knowledge. This "linkage" is habitually associated with the operation of conjunction and thus, in the simplest cases, later background knowledge assumes the form $k \& p$, or $k \& \sim p$. As regards this "expansion" of background knowledge, we must understand that while the contents of knowledge increase, its extension remains unchanged or even diminished, as $k \& p$ or $k \& \sim p$ may correspond to an area of reality at most as large as k.

It may be questioned whether p or $\sim p$ must be linked to k necessarily in conjunction. If we seek an answer merely from an epistemological point of view, we can suggest the following. There are cognitive situations which bring about an increase in our knowledge not only in content but in extension as well. In such cases we generally know not only more but something else than before. A cognitive situation like this is called a revolution. The more or less radical transformation of the background knowledge can be represented by the formulas $k \vee p$, or $k \vee \sim p$. At this point, however, there arise certain logical difficulties to which we shall return later.

The results of our investigation can be summed up as follows. As the object of argumentation, the hypothesis h may, on the basis of the initial background knowledge k, ascribe a certain (perhaps maximum) probability to the statements p, or $\sim p$. Then these may be resolved at the second stage of the diachronic approach, owing to which the latter will no longer be hypotheses but part of the later background of knowledge and will be added to the store of true knowledge. This will be so even if the additional knowledge as a hypothesis is refuted, for its negation is true, and so our knowledge has increased.

An important condition is for *the earlier and the later background of knowledge to be comparable.* Let k_i stand for the earlier, and k_j for the later background of knowledge. The background knowledge k_i is comparable with k_j if, and only if, k_i is a logical consequence of k_j i.e., $k_j \Rightarrow k_i$, where $k \geq i$, and assumes the values of the indices of the following graph:

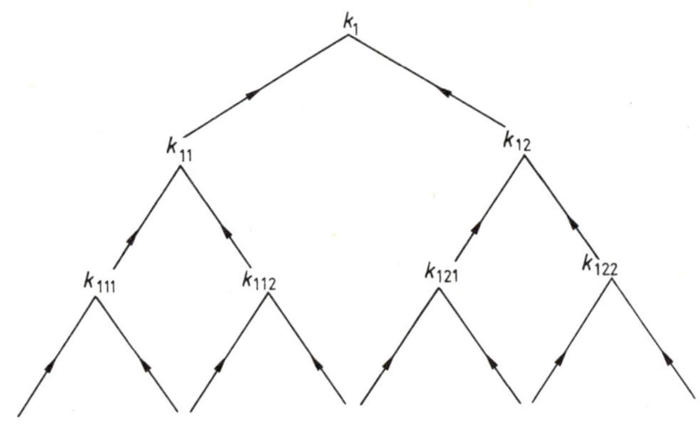

Figure 1

The graph shows that the backgrounds of knowledge which appear on the same branch are only comparable. E.g., k_{112} is comparable with k_{11} and k_1, but it is not comparable with k_{111}, k_{121}, k_{122}, and k_{12}. Since $k_1 \equiv (k_1 \,\&\, p) \lor (k_1 \,\&\, \sim p) \equiv (k_1 \,\&\, p \,\&\, q) \lor (k_1 \,\&\, p \,\&\, \sim q) \lor \lor (k_1 \,\&\, p \,\&\, q) \lor (k_1 \,\&\, \sim p \,\&\, \sim q)$, we get the following graph:

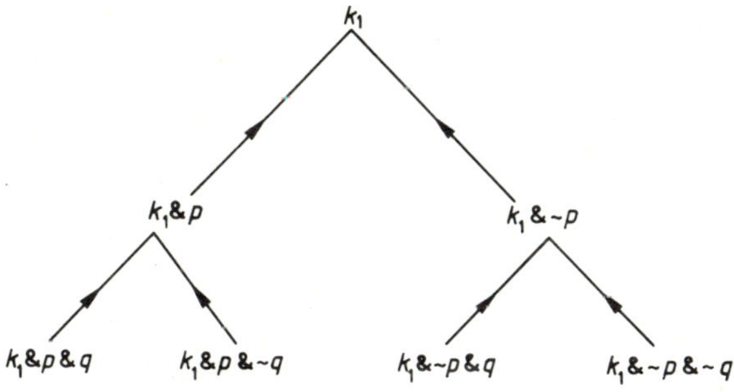

Figure 2

Figure 2 describes the process of the expansion of background knowledge. The initial background knowledge as a conjunctive component is common in all the later backgrounds of knowledge and is therefore comparable with each. This rather simple logical model of the cumulative growth of knowledge will of course become inadequate as soon as cognition undergoes a more or less radical change.

At this stage the hypotheses p or $\sim p$ cease to function as mere additional knowledge and their eventual verification gives them such a great importance that they replace the initial background knowledge k. The new background knowledge p will then only incorporate that fragment of k which is compatible with it, i.e., the new background knowledge will have a component $k \,\&\, p$. This will represent the part of the studied field which is the shared element of continuity in both the old and the new knowledge. Since the extension of p is wider than that of $k \,\&\, p$, a component $\sim k \,\&\, p$ must also be taken into account, which is connected through alternation to $k \,\&\, p$ according to the equivalence $p \equiv (k \,\&\, p) \lor (\sim k \,\&\, p)$. Both $k \,\&\, p$ and $\sim k \,\&\, p$ are obviously comparable with p, for $k \,\&\, p \Rightarrow p$; and $\sim k \,\&\, p \Rightarrow p$.

On the analogy of Figure 2, comparability can be represented as follows:

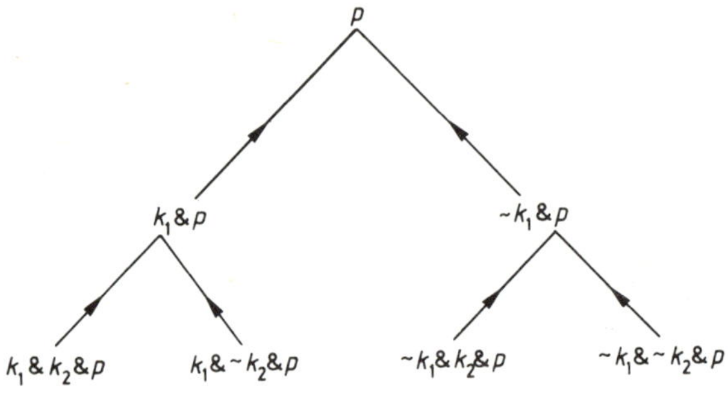

Figure 3

where k_1 is the background knowledge preceding the latest revolutionary situation and k_2 is the one before the revolution bringing about k_1. In retrospect, it appears that the remaining fragments of earlier backgrounds of knowledge may lay among cognitive elements the strongest claim to absolute truth. These, however, can only be identified post festa, which is why a contemporary cannot be certain about the truth of k_2 and/or k_1 just the same way as we cannot know the truth of p for certain. Consequently e.g. the branch k_1 & k_2 & $p \Rightarrow k_1$ & $p \Rightarrow p$ of the graph does not describe the actual process of the development of knowledge but *its reconstruction based on the present stage of knowledge,* the structure of which reconstruction is the same as that of a diachronic situation of reasoning. (See Ch. 5.)

A closer look at the graph may reveal that even after revolutionary changes in knowledge there remains a certain degree of continuity, for the new background knowledge p always contains a fragment of the old background knowledge k. However radical a scientific revolution may be, the continuity of human knowledge is never broken. Yet on a given level of cognition it is impossible to tell with certainty which fragment of knowledge will maintain the continuity. Earlier knowledge is frequently embedded in the new background knowledge in a way that is not mechanical but involves the transformation of the prior knowledge,

28

too. Let that transformed prior knowledge be designated with a k'. Then it obtains that $p \Rightarrow k'$ (which does not preclude the truth of $k' \& p \Rightarrow p$).

We are well aware of the insufficiency of this treatment of comparability but it may temporarily satisfy our needs for some foundation of the investigations to come. (For a more detailed discussion of the issue see Appendix 4.)

In scientific literature, qualifications like "at the present state of our knowledge" or "in the light of latest information" are often encountered. Such phrases may be interpreted as global references to the background knowledge. Explicit reference, however, is most usually made only of those arguments within the background knowledge which are ascribed an especially important role or appear in the inferences used in the argumentation. This is why it is necessary to separate the additional knowledge within the posterior background knowledge.

Since we are examining rational, not actual, situations of reasoning, both the prior and the posterior background knowledge are taken to represent the *world level of relevant knowledge at any given time* (that is, the status quo of a relatively well defined, homogeneous area of human knowledge).

Given the fact that backgrounds of knowledge are complexes of specific scientific knowledge deemed relevant and true, they bear the same marks as determine the place of a science in the system of sciences. The so-called abstract or formal sciences, like symbolic logic and mathematics, have backgrounds of knowledge that only contain theoretical knowledge. However, the backgrounds of knowledge of the factual sciences (the natural and social sciences) always include empirical knowledge (data, descriptions, rules obtained through empirical generalization, etc.) as well. This is especially true of the additional knowledge acquired in order to improve the grounding of hypotheses.

As it has been mentioned, the background knowledge k is relative since it reflects the historical state of development of a given science. But at the same time it is a basis of comparison, the logical foundation for the evaluation of hypotheses. Therefore, on a given level of cognition it can be considered as absolute.

4 RELATIVE TRUTH

It has already been made clear that hypotheses cannot be compared with the states of affairs assumed to be correlated with them except through mediation by background knowledge. Therefore we can only speak of an indirect correspondence between hypotheses and reality, which always involves a reference to the background knowledge as a factor introducing relativity. The truth value of a hypothesis is then a relation of adequacy mediated by the truth of its background knowledge and by the relation of corroboration obtaining between the background knowledge as a complex of arguments and the hypothesis. The line of the argument is like this: given that the background knowledge is true and it grounds the hypothesis, the hypothesis is itself true with reference to, or on the basis of, the background knowledge.

The truth of the background knowledge as a basis of reference is regarded as absolute at a given stage of cognition. Hypotheses are also assumed to have such absolute truth values attached to them in their relation to reality. However, this truth value cannot directly be defined, it can only be reconstructed in an intricate logical way. The reconstruction is effected through rational argumentation which, in its turn, requires that the concept of the truth value of a hypothesis relative to its background knowledge be introduced. We shall briefly term this concept *relative truth value.*

There are degrees of the relative truth values of hypotheses. In many cases we cannot be content with labels like "relatively true" or "relatively false": it is necessary to find more adequate qualifications. This may be accomplished by introducing a continuous scale of values to measure the degrees of relative truth. The scale must have the maximal and the minimal values fixed, which make it possible to establish the so-called intervening values, and to arrange them in rank order.

For the moment we are not considering the question whether adequate numerical values can be assigned to the degrees of relative truth and, if so, within what interval these values may occur.

If the relative truth value of a hypothesis is maximal, it is qualified as *true for certain*, or, in short, *certain*. Note that hypotheses, when they are certain, cannot in fact count as hypotheses any more since the information they embody is so reliable that they can just as well be looked upon as background knowledge. Still, as extreme cases, it is reasonable to consider them to be hypotheses once it is their truth value that is taken to be maximal. Certainly true are e.g. tautologies in classical two-valued logic, or proved theorems in mathematics, verified nomological or theoretical hypotheses in natural or social sciences, scientifically valid data, etc.

The minimum degree of relative truth is represented by the truth values of hypotheses which *cannot be true*, i.e. which are *precluded* or *impossible*. Logical contradictions, or the negations of mathematical theorems, scientific laws, or theories are said to be impossible. Once again, hypotheses of the truth value "impossible" are not "real" hypotheses but borderline cases possessing a value diagonally opposite to that of certainly true propositions.

If the truth of a hypothesis is not precluded, its relative truth value is "*possibly true*" or simply "*possible*". Hypotheses which are not true for certain are classified under the category "*non-certain*". Hypotheses that are neither certain nor impossible make up the class of *contingently true* hypotheses. These we briefly call contingent. The truth value of a real scientific hypothesis is always contingent.

A specific place on the scale of relative truth values is occupied by hypotheses which fall approximately in the middle between the "precluded" and the "certain". These are said to be *half true* and to have the status of *half-truth*.

Hypotheses of a relative truth value bigger than half true are termed *probably true*, while those of a value smaller than that are said to be *probably false*. If a hypothesis is actually true even though not certain, it is *eventually true*, whereas if it is actually false, though not impossible, it is *eventually false*. The truth values listed above are altogether called truth modalities.

The scale of relative truth values is represented by the following diagram:

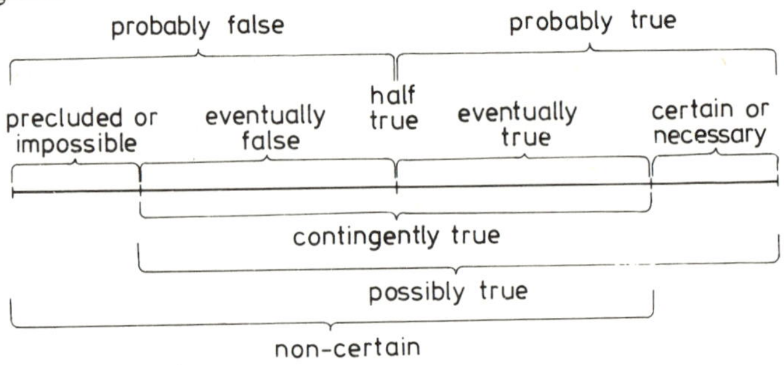

Figure 4

It is easy to recognize that the concept of relative truth is a generalization of the classical truth concept. This is seen if we omit all truth modalities except "probably true" and "probably false". The modifier "probably" will not be needed any more. Then, if we further omit any reference to background knowledge, the distinction between "relative truth" and "truth" will be dispensable.

In synchronical reasoning situations, relative truth values are always referred to the same background of knowledge, therefore most of the time the omission of the reference does not lead to misunderstanding. In diachronical argumentation, there are at least two backgrounds of knowledge used as frames of reference and, consequently, it is necessary to differentiate between *initial* and (following revision) *later truth values*. Nevertheless in most cases there is no ambiguity if reference is made to the later background of knowledge only.

In what follows, the relative truth value of hypotheses will often be called their *plausibility*, taking the term in the sense used by Polya (Polya, 1954, II, 112—115). A similar conception is held by Russell, who speaks of "credibility" (Russell, 1948, 398—414, and 506—512).

Knowing certain plausibilities, on this basis we can determine unknown plausibilities through logical operations. These thought processes are called *plausible inferences*.

5 MAIN TYPES OF PLAUSIBLE INFERENCES

Let k continue to designate the initial background knowledge, and $h_1, h_2, \ldots$ the hypotheses forming the theses of argumentation. Hypotheses that have for both their premisses and their conclusions the relative truth values of certain hypotheses measured against comparable backgrounds of knowledge will be called *plausible inferences*. Below, an informal construction of the theory of plausible inferences is put forward.

Let us consider the following plausible inferences, in the propositions of which the background knowledge k is naturally presupposed:

$$
\begin{array}{lll}
S1 & h_1 \supset h_2 & \text{is certain} \\
   & h_1 & \text{is certain} \\
\hline
   & h_2 & \text{is certain} \\
S'1 & h_1 \supset h_2 & \text{is certain} \\
   & h_1 & \text{is possible} \\
\hline
\end{array}
$$

h_2 is at least as plausible as h_1.

$S1$ and $S'1$ are *synchronic plausible inferences* since the frame of reference for their premisses and conclusions is always the background knowledge k. Their truth can be seen intuitively, for their correspondence to modus ponens is plausible. Precise formulation and proof will be left for later.

If the conclusion of a plausible inference contains either the maximum (certainly true) or the minimum (precluded from being true) truth value only, we are dealing with a *strong* plausible inference, otherwise with a *weak* one. Thus $S1$ is a strong, while $S'1$ is a weak (synchronic) plausible inference.

In a similar fashion, we can supply the plausible correlatives of modus tollens, modus tollendo ponens, and modus ponendo tollens. References to the background knowledge k will be omitted for simplicity's sake.

$S2$	$h_1 \supset h_2$	certain		$S'2$	$h_1 \supset h_2$	certain
	$\sim h_2$	certain			$\sim h_2$	possible
	$\sim h_1$	certain			$\sim h_1$ at least as plausible as $\sim h_2$	

$S3$	$h_1 \vee h_2$	certain		$S'3$	$h_1 \vee h_2$	certain
	$\sim h_1$	certain			$\sim h_1$	possible
	h_2	certain			h_2 at least as plausible as $\sim h_1$	

$S4$	$h_1 \mid h_2$	certain		$S'4$	$h_1 \mid h_2$	certain
	h_1	certain			h_1	possible
	$\sim h_2$	certain			$\sim h_2$ at least as plausible as h_1.	

Furthermore, let us supply the correlatives of the hypothetical syllogism:

$S5$	$h_1 \supset h_2$	certain		$S'5$	$h_1 \supset h_2$	certain
	$h_2 \supset h_3$	certain			$h_2 \supset h_3$	more plausible than h_3
	$h_1 \supset h_3$	certain			$h_1 \supset h_3$	more plausible than h_3

In scientific reasoning, a considerably greater part is played by *diachronic plausible inferences* than by synchronic ones, and the former are also richer in variety of form than the latter.

34

The difference between strong (D) and weak (D') plausible inferences will be made here, too, and references to the background knowledge k will be omitted as above.

$D1$ $h_1 \supset h_2$ is certain
      h_1 is possible
____
      h_2 is certain assumed
        that h_1

$D'1$ $h_1 \supset h_2$ is certain
      h_1 is possible
      h_2 is non-certain
____
      h_1 is more plausible
        assumed that h_2

$D2$ $h_1 \supset h_2$ is certain
      $\sim h_2$ is possible
____
      $\sim h_1$ is certain assumed
        that $\sim h_2$

$D'2$ $h_1 \supset h_2$ is certain
      $\sim h_1$ is non-certain
      $\sim h_2$ is possible
____
      $\sim h_2$ is more plausible
        assumed that $\sim h_1$

$D3$ $h_1 \vee h_2$ is certain
      $\sim h_1$ is possible
____
      h_2 is certain assumed
        that $\sim h_1$

$D'3$ $h_1 \vee h_2$ is certain
      $\sim h_1$ is possible
      h_2 is non-certain
____
      $\sim h_1$ is more plausible
        assumed that h_2

$D4$ $h_1 | h_2$ is certain
      h_1 is possible
____
      $\sim h_2$ is certain assumed
        that h_1

$D'4$ $h_1 | h_2$ is certain
      h_1 is possible
      $\sim h_2$ is non-certain
____
      h_1 is more plausible
        assumed that $\sim h_2$.

The conclusions of $D'1$—$D'4$ need some clarification. The formula ". . . more plausible assumed that. . . " is an abbreviation of ". . . more plausible on the basis of the background knowledge k, and of the proof of. . . , than on the basis of the background knowledge k alone".

$D5$ $h_1 \supset h_2$ is certain
      $h_2 \supset h_3$ is certain
      h_1 is possible
____

   h_3 is certain assumed
      that h_1

$D'5.1$ $h_1 \supset h_2$ is certain
       $h_1 \supset h_3$ is certain
       h_1 is possible
       $h_2 \vee h_3$ is non-certain
the plausibility of h_3, when $\sim h_1$, is not decreased by the proof of the validity of h_2
____

   h_3 is more plausible
      assumed that h_2

$D'5.2$ $h_1 \supset h_2$ is certain
       $h_1 \supset h_3$ is certain
       h_1 is possible
       $h_2 \vee h_3$ is not more plausible than $\sim(h_2 \vee h_3)$
       $h_2 \,\&\, h_3$ is not more plausible than $\sim(h_2 \,\&\, h_3)$
____

   h_2 is less plausible assumed
      that $\sim h_3$.

The intuitive recognition of the validity of $D'5.1$ may be facilitated by the following example:

Certain:	If light is of a wave nature, it has the property of interference.
Certain:	If light is of a wave nature, it has the property of diffraction.
Possible:	Light is of a wave nature.
Uncertain:	Light has the properties of interference and/or diffraction.

Should it turn out that light is not of a wave nature, the eventual establishment of light interference should nevertheless not decrease the plausibility of its property of diffraction.

The interference of light having been established, its property of diffraction is more plausible.

$D'5.1$ and $D'5.2$ are specific cases of inference by analogy.

Below, some weak diachronic rules of inference are further mentioned, representing the reasoning situation of *experimentum crucis:*

$S'6$

$h_1 \supset h_2$	is certain
$h_3 \supset h_2$	is non-certain
h_1 and h_3 are equally plausible	

---

h_1 is more plausible on the basis of h_2 than h_3 on the basis of h_2

$S'7$

$h_1 \supset h_2$	is certain
$h_3 \supset \sim h_2$	is certain
h_1	is possible
h_3	is possible
h_2	is non-certain

---

h_1 is impossible and h_3 is more plausible on the basis of $\sim h_2$

$S'8$

$h_1 \mid h_3$	is certain
$h_1 \supset h_2$	is certain
$\sim h_3 \supset h_2$	is certain
h_1 and h_3 are possible and equally plausible	
h_2 is non-certain	

---

on the basis of h_2, the plausibility of h_1 increases while the plausibility of h_3 decreases, and the plausibility of h_1 on the basis of h_2 is greater than the plausibility of h_3 on the basis of h_2

$S'9$

$h_1 \mid h_3$	is certain
$h_1 \supset h_2$	is certain
$\sim h_3 \supset h_2$	is certain
h_1 and h_3 are possible and equally plausible	
h_2 is non-certain	

---

h_1 is impossible and h_3 is certain on the basis of $\sim h_2$.

The above plausible inferences are just a few examples of the forms of ratiocination used in scientific reasoning. All of these inferences can be discussed in the so-called comparative probabilistic logic.

6 COMPARATIVE PROBABILISTIC LOGIC

This chapter is designed to offer a sketch of probabilistic logic, a means expected to be fit for an exact formulation and proof of the above outlined plausible inferences.

The fundamental concept of probabilistic logic is the function "$P(p, k)$", in which p is some proposition (hypothesis), and k is the previously mentioned background knowledge. The latter in every case is assumed to be self-consistent. The propositions p and k come under the laws of classical two-valued logic. The function P expresses the probability of p with regard to k. Its possible values may be real numbers within the closed interval 0 to 1, among which the arithmetical relations "$=$", "$<$", and "$\leq$", are interpreted. This, however, does not necessitate numerical interpretation. E.g. the non-numerical interpretation of $P(p, k) < P(q, k)$ runs like this: on the basis of k, q is more probable than k. If any change in the background knowledge were precluded, the reference to k would be superfluous and "$P(p)$" would be enough. However, probabilistic logic must account for possible changes in the background knowledge as well. Accordingly laws establishing relations between $P(p, k)$ and $P(p, k \mathbin{\&} q)$ will be encountered later.

Probabilistic logic has the task of exploring relations concerning the function P. For this purpose, laws of classical two-valued logic and of the arithmetic of real numbers are used as well as the following axioms.

(A1) $\qquad 0 \leq P(p, k) \leq 1$.

(A2) $\qquad P(p, k) + P(\sim p, k) = 1$.

(A3) $P(p \& q, k) = P(p, k) \cdot P(q, k \& p)$

provided $k \& p$ is not logically false.

(A3')   If $P(p, k) \neq 0$, then $\dfrac{P(p \& q, k)}{P(p, k)} = P(q, k \& p)$.

(A4) $P(k, k) = 1$.

(A5)   The value of the function $P(p, k)$ remains unchanged if either p or k is replaced by a logically equivalent statement. (A5) detailed:

(A5.1)   If $p \Leftrightarrow q$, then $P(p, k) = P(q, k)$.

(A5.2)   If $k_1 \Leftrightarrow k_2$, then $P(p, k_1) = P(p, k_2)$:

These axioms essentially correspond to the postulates of G. H. von Wright's abstract probabilistic calculus (von Wright, 1957, 91—96).

Logical probability will hereafter be considered to be the logical explication of plausibility, i.e. of relative truth. Here are the *non-numerical interpretations* of some relations:

"$P(p, k) = 1$"	means	"p is certain on the basis of k" ("the plausibility of p on the basis of k is maximal").
"$P(p, k) = 0$"	means	"p is precluded on the basis of k" ("the plausibility of p on the basis of k is minimal").
"$P(p, k) < 1$"	means	"p is not certain on the basis of k" ("the plausibility of p on the basis of k is not maximal").
"$P(p, k) > 0$"	means	"p is not precluded on the basis of k", or "p is possible on the basis of k" ("the plausibility of p on the basis of k is not minimal").
"$0 < P(p, k) < 1$"	means	"p is contingent on the basis of k" ("the plausibility of p on the basis of k is neither maximal nor minimal").

"$P(p, k) = P(\sim p, k)$"	means	"p is undecided on the basis of k", or "p is half true on the basis of k" ("on the basis of k, p is as plausible as its negation").
"$P(p, k) = P(q, k)$"	means	"on the basis of k, p and q are equally plausible".
$P(p, k) < P(q, k)$"	means	"on the basis of k, q is more plausible than p".
"$P(p, k) > P(\sim p, k)$"	means	"p is plausible on the basis of k".
"$P(p, k) \leqq P(q, k)$"	means	"on the basis of k, p is not more plausible than q", i.e. "q is at least as plausible as p".
"$P(p, k) < P(p, k \,\&\, q)$"	means	"p is more plausible on the basis of k and q than on the basis of k" i.e. "the justification of q increases the plausibility of p".

Proceeding from these examples, the reader may construct the non-numerical interpretations of other, similar relations too.

Now we are going to put forward the main theorems of probabilistic logic, some of the simplest without proofs:

T1   If t is a logical truth, then $P(t, k) = 1$.

T2   If p is a consequence of k, i.e. $k \Rightarrow p$, then $P(p, k) = 1$.

T3   If q is a consequence of p, then $P(p \supset q, k) = 1$.

T4   If f is logically false, then $P(f, k) = 0$.

T5   If $\sim p$ is a consequence of k, then $P(p, k) = 0$.

T6   If $\sim q$ is a consequence of p, then $P(p \supset q, k) = 0$.

T7   If p is a consequence of k, then $P(q, k) = P(q, k \,\&\, p)$.

T8 $P(p, k) = P(p \,\&\, q, k) + P(p \,\&\, \sim q, k)$.

T9 $P(p \vee q, k) = P(p, k) + P(q, k) - P(p \,\&\, q, k)$.

T10   If $P(p \,\&\, q, k) = 0$, then $P(p \vee q, k) = P(p, k) + P(q, k)$.

T11 $\qquad P(p\ \&\ q, k) \leq P(p, k) \leq P(p \vee q, k)$.

T12 $\qquad P(p \supset q, k) = P(\sim p, k) + P(p\ \&\ q, k)$.

T13 $\qquad$ If $P(p \supset q, k) = 1$, then $P(p, k) = P(p\ \&\ q, k)$ and
$P(p\ \&\ \sim q, k) = 0$.

T14 $\qquad$ If $P(p \supset q, k) = 1$, then $P(p, k) \leq P(q, k)$.

T15 $\qquad$ If $P(p \supset q, k) = 1$, and $P(p, k) = 1$, then $P(q, k) = 1$.

T16 $\qquad$ If $\{p_1, \ldots, p_n\} \Rightarrow q$, then $\{P(p_1, k) = 1 \ldots, P(p_n, k) = 1\}$
$\Rightarrow P(q, k) = 1$.
$$n \geq 1$$

T17 $\qquad$ If $P(p \equiv q, k) = 1$, then $P(p, k) = P(q, k)$.

T18 $\qquad$ If $k \Rightarrow p \vee q$, $k \Rightarrow p|q$, and $P(p, k) = P(q, k)$, then $P(p, k) = 1/2$
and $P(q, k) = 1/2$.

Proof:

(1) $\qquad$ By virtue of T2, it follows from the first two premisses that
$P(p \vee q, k) = 1$ and $P(p|q, k) = 1$.

(2) $\qquad$ From (1), applying T10:
$P(p, k) + P(q, k) = 1$.

(3) $\qquad$ (2) and the third premiss directly yield the theorem to be
proved.

T19 $\qquad$ If $k \Rightarrow (p \vee q)$ and $k \Rightarrow p|q$ and $P(p, k) \geq P(q, k)$ then
$P(p, k) \geq 1/2 \geq P(q, k)$.
The proof of this theorem is analogous to that of the
preceding one.

T20 $\qquad$ If $k \Rightarrow q$, then $P(p, k\ \&\ q) = P(p, k)$ i.e. q which is certain on
the basis of k is irrelevant to p when k.

Proof:

(1) $\qquad P(p, k) = P(q, k)P(p, k\ \&\ q) + P(\sim q, k)P(p, k\ \&\ \sim q)$.

(2) $\qquad$ Under the given condition
$P(q, k) = 1$ and $P(\sim q, k) = 0$.

(3)   According to (1) and (2),
$$P(p, k \ \& \ q) = P(p, k) \ .$$

T21   If $k \Rightarrow \sim q$, then $P(p, k \ \& \sim q) = P(p, k)$, i.e. q precluded on the basis of k is irrelevant to p when k.
The proof is the same as in T20.

T22   If $a)\ k \Rightarrow p$ and $b)\ k \nRightarrow \sim q$, then $P(p, k \ \& \ q) = P(p, k)$, i.e. every q that is not precluded on the basis of k is irrelevant to some p which is certain on the basis of k.

Proof:

(1)   On the basis of T21
$$P(q, k) = P(q, k \ \& \ p) \ .$$

(2)   Since
$$P(q, k)P(p, k \ \& \ q) = P(p, k)P(q, k \ \& \ p),$$

(3)   therefore
$$P(p, k \ \& \ q) = P(p, k) \ .$$

T23   If $a)\ k \Rightarrow \sim p$ and $b)\ k \nRightarrow \sim q$ then $P(p, k \ \& \ q) = P(p, k)$, i.e. every q that is not precluded on the basis of k is irrelevant to some p precluded on the basis of k.

Proof:

(1) $P(p, k) = 0$ in virtue of condition $a)$ and $P(q, k) > 0$ in virtue of condition $b)$.

(2) $P(p, k) = P(q, k)P(p, k \ \& \ q) + P(\sim q, k)P(p, k \ \& \sim q) \ .$

(3)   In accordance with (1) and (2),
$$P(p, k) = P(p, k)P(p, k \ \& \ q) = 0.$$

(4)   Since $P(q, k) > 0$, therefore $P(p, k) = P(p, k \ \& \ q) = 0.$

T24   If $P(p, k \ \& \ q) > P(p, k)$, then $P(q, k \ \& \ p) > P(q, k)$, i.e. if q is positively relevant to p when k, then p is also positively relevant to q when k.

T25   If $P(p, k \& q) < P(p, k)$, then $P(q, k \& p) < P(q, k)$, i.e. if q is negatively relevant to p when k, then p is negatively relevant to q.

T26   If $P(h, k \& p) > P(h, k)$, then $P(h, k \& \sim q) < P(h, k)$, i.e. if p is positively relevant to h when k, then $\sim p$ is negatively relevant to h.

Proof:

(1)   Since
$$P(h, k \& p) > P(h, k),$$
$$P(h \& p, k) > P(h, k)P(p, k).$$

(2)   Subtract
$$P(h \& p, k)P(h \vee p, k)$$
from both sides of the inequality.

(3)   The substitution of $P(h \vee p, k) = P(h, k) + P(p, k) - P(h \& p, k)$ results $P(h \& p, k) - P(h, k)P(h \& p, k) - P(p, k)P(h \& p, k) + P(h \& p, k)^2 > P(p, k)P(h, k) - P(h, k)P(h \& p, k) - P(p, k)P(h \& p, k) + P(h \& p, k)^2$.

(4)   Hence
$$P(h \& p, k)(1 - P(h, k) - P(p, k) + P(h \& p, k)) > (P(h, k) - P(h \& p, k))(P(p, k) - P(h \& p, k)).$$

(5)   Considering that $1 - P(h, k) - P(p, k) + P(h \& p, k) = P(\sim h \& \sim p, k)$, $P(h, k) - P(h \& p, k) = P(h \& \sim p, k)$ and $P(p, k) - P(h \& p, k) = P(\sim h \& p, k)$ it follows that
$$P(h \& p, k)P(\sim h \& \sim p, k) > P(\sim h \& p, k)P(h \& \sim p, k).$$

(6)   On the basis of $P(\sim h \& p, k) = 1 - P(h, k) - P(\sim p, k) + P(h \& \sim p)$, $P(h \& p, k) = P(h, k) - P(h \& \sim p, k)$ and $P(\sim h \& \sim p, k) = P(\sim p, k) - P(h \& \sim p, k)$, substitution and multiplication results that $P(h \& \sim p, k) - P(h, k)P(h \& \sim p, k) - P(\sim p, k)P(h \& \sim p, k) + P(h \& \sim p, k)^2 < P(h, k)P(\sim p, k) - P(\sim p, k)P(h \& \sim p, k) - P(h, k)P(h \& \sim p, k) + P(h \& \sim p, k)^2$.

(7)   After reduction,
$$P(h \mathbin{\&} \sim p, k) < P(h, k)P(\sim p, k).$$

(8)   From (7)
$$P(h, k \mathbin{\&} \sim p) < P(h, k).$$

T27   If $P(h, k \mathbin{\&} p) = P(h, k)$, then $P(h, k \mathbin{\&} \sim p) = P(h, k)$, i.e. if p is irrelevant to h when k then p is also irrelevant to h.

The proof is the same as in T26.

Now we are turning to *probabilistic inferences* regarded as exact formulations of the above plausible inferences. By a probabilistic inference is meant an inference the premisses and conclusion of which contain arithmetical statements about the probabilistic measure function P and the conclusion logically follows (in the sense of classical two-valued logic) from the premisses, from axioms (A1)—(A5) of probabilistic logic, and from the fundamental axioms of the arithmetic of real numbers. In this way (in the sense of classical two-valued logic) probabilistic inferences are *enthymematic inferences* whose implicit premisses are taken from among the axioms of probabilistic logic and real arithmetic. Having accepted the implicit axioms besides the premisses, the conclusion can no more be doubted than in deductive inferences.

It may not be necessary to formulate and prove the analogues V_S and V_D of every plausible inference of the types S and D. Let it suffice to put forward a few characteristic examples to demonstrate the efficiency of probabilistic logic in the explication of inferences.

$V_S 1$ $\{P(h_1 \supset h_2, k) = 1,\ P(h_1, k) = 1\} \Rightarrow P(h_2, k) = 1.$

$V_S' 1$ $\{P(h_1 \supset h_2, k) = 1,\ P(h_1, k) > 0\} \Rightarrow 0 < P(h_1, k) \leq P(h_2, k).$

       Proof: $V_S 1$ is a transformation of T16, and $V_S' 1$ is one of T14.

$V_S 5$ $\{P(h_1 \supset h_2, k) = 1,\ P(h_2 \supset h_3, k) = 1\} \Rightarrow P(h_1 \supset h_3, k) = 1$

       Proof: According to T16.

V'_S5 $\{P(h_1 \supset h_2, k)=1, \quad P(h_3, k) < P(h_2 \supset h_3, k)\} \Rightarrow P(h_3, k) < < P(h_1 \supset h_3, k)$

Proof:

(1) Since $(h_1 \supset h_2) \Rightarrow [(h_1 \supset h_3) \equiv (h_1 \,\&\, h_2) \supset h_3]$, therefore, according to T2, $P[((h_1 \supset h_3) \equiv (h_1 \,\&\, h_2) \supset h_3] = 1$.

(2) Hence, according to T17 $P[(h_1 \,\&\, h_2) \supset h_3, k] = = P(h_1 \supset h_3, k)$.

(3) Consider that
$$P[(h_2 \supset h_3) \supset (h_1 \,\&\, h_2) \supset h_3, k] = 1.$$

(4) Then, in virtue of T14, it follows that
$$P(h_2 \supset h_3, k) \leq P[(h_1 \,\&\, h_2) \supset h_3, k].$$

(5) But according to the second premiss,
$$P(h_3, k) < P(h_2 \supset h_3, k).$$

(6) Reading off theses (5), (4), and (3) together (in that order) now we have the thesis to be proved.

V_D1 $\{P(h_1 \supset h_2, k)=1, P(h_1, k) > 0 \Rightarrow P(h_2, k \,\&\, h_1) = 1.$

Proof:

(1) On the basis of the first premiss and T13
$$P(h_1, k) = P(h_1 \,\&\, h_2, k).$$

(2) According to (A3′)
$$P(h_2, k \,\&\, h_1) = \frac{P(h_1 \,\&\, h_2, k)}{P(h_1, k)}.$$

(3) From (1) and (2), making use of the second premiss:
$$P(h_2, k \,\&\, h_1) = \frac{P(h_1, k)}{P(h_1, k)} = 1.$$

V'_D1 $\{P(h_1 \supset h_2, k)=1, P(h_1 k) > 0, P(h_2, k) < 1\} \Rightarrow P(h_1 k \,\&\, h_2) > > P(h_1, k).$

Proof:

(1) On the basis of the first premiss, according to T13
$$P(h_1, k) = P(h_1 \,\&\, h_2, k).$$

(2) According to (A3′)
$$P(h_1, k \,\&\, h_2) \cdot P(h_2, k) = P(h_1, k).$$

(3) From the premisses and using T14 we obtain that
$$0 < P(h_2, k) < 1.$$

(4) On the basis of (2) and (3)
$$P(h_1, k \,\&\, h_2) > P(h_1, k).$$

$V'_D 5.1$ $\{P(h_1 \supset h_2, k) = 1,\ P(h_1 \supset h_3, k) = 1,\ P(h_1, k) > 0,$
$P(h_2 \vee h_3, k) < 1,\ P(h_3, k \,\&\, \sim h_1) \le$
$\le P(h_3, k \,\&\, \sim h_1 \,\&\, h_2)\} \Rightarrow P(h_3, k \,\&\, h_2) > P(h_3, k).$

Proof: To provide a synoptical proof of $V'_D 5$, it will be convenient to use Venn Diagrams. Let probability 1 be represented by a square, which is in fact a measure of the background knowledge k related to itself (cf. axiom (A5)). The measures of the hypotheses h_1, h_2 and h_3 are represented by intersecting circles within the square. Each of the resulting 8 elementary areas will correspond to a value $P(\pm h_1 \,\&\, \pm h_2 \,\&\, \pm h_3, k)$, where the place of the sign "$\pm$" is taken by the sign of negation, or by nothing.

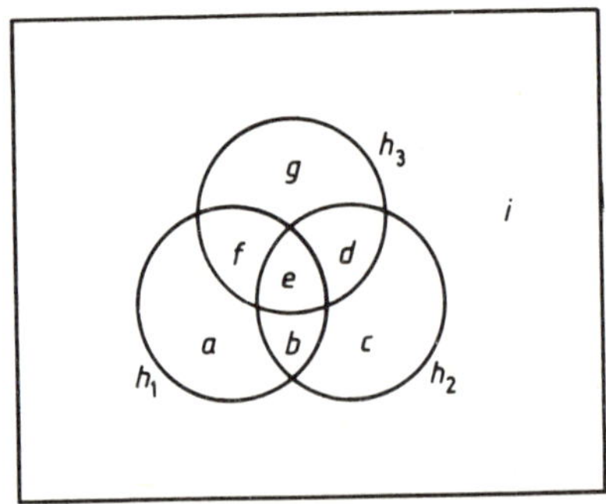

a: $P(h_1 \,\&\, \sim h_2 \,\&\, \sim h_3, k)$
b: $P(h_1 \,\&\, h_2 \,\&\, \sim h_3, k)$
c: $P(\sim h_1 \,\&\, h_2 \,\&\, \sim h_3, k)$
d: $P(\sim h_1 \,\&\, h_2 \,\&\, h_3, k)$
e: $P(h_1 \,\&\, h_2 \,\&\, h_3, k)$
f: $P(h_1 \,\&\, \sim h_2 \,\&\, h_3, k)$
g: $P(\sim h_1 \,\&\, \sim h_2 \,\&\, h_3, k)$
i: $P(\sim h_1 \,\&\, \sim h_2 \,\&\, \sim h_3, k)$

Figure 5

(1)   It is readily seen that
$$a + b + c + d + e + f + g + i = 1.$$

Note: Part of our information can be filled in the empty diagram. E.g. "$P(h_1 \,\&\, h_2, k) = 0$" means that the probabilistic measure of the overlapping parts of the circles h_1 and h_2 is 0. This is marked by shading out the overlapping. According to the first and the second premiss the diagram will be like this:

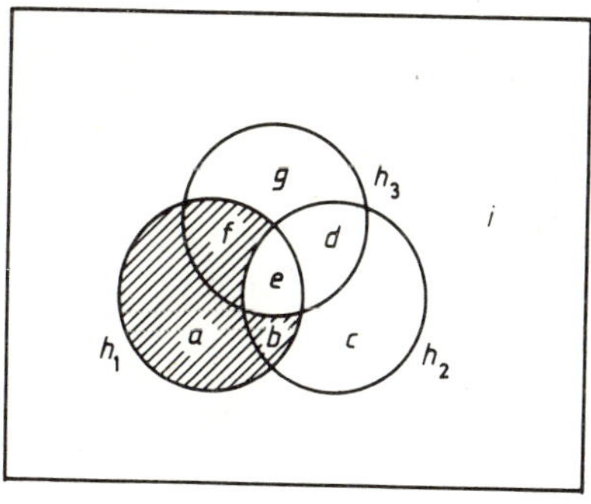

Figure 6

(2)   If
$$c + d > 0,$$

(3)   then the fifth premiss can be expressed like this:
$$\frac{d+g}{c+d+g+i} \leq \frac{d}{c+d}.$$

(4)   According to the third and the fourth premiss,
$$e > 0$$

(5) $c + d + e + g < 1$, that is, $i > 0$.

(6)   A little calculation from (3) will produce that $cg \leq di$.

(7)   According to (4) and (5), $ei > 0$. Also, $c + d + e + g + i = 1$. From these it follows that $(c+d+e+g+i)(d+e) - ei < de$.

(8)   This, on the strength of (6), yields that
$$(c+d+e)(d+e+g)<d+e,$$
that is
$$d+e+g < \frac{d+e}{c+d+e}.$$

(9)   If (2) is not fulfilled, i.e. $c+d=0$ (when in fact $d=0$), the latter inequality takes the form
$$e+g < \frac{e}{e} = 1.$$

(10)   Otherwise
$$d+e+g < \frac{d+e}{c+d+e} < 1.$$

Since $P(h_3,k)=d+e+g$, $P(h_2 \& h_3,k)=d+e$, and $P(h_2,k)=c+d+e$, inequality (10) corroborates the conclusion of $V'_D 5.1$.

$V'_D 5.2$ $\{P(h_1 \supset h_2,k)=1,\ P(h_1 \supset h_3,k)=1,\ P(h,k)>0,$
$P(h_2 \vee h_3,k) \leq P(\sim(h_2 \vee h_3),k),\ P(h_2 \& h_3,k) \leq$
$\leq P(h_2|h_3,k)\} \Rightarrow P(h_3,k) > P(h_3,k \& \sim h_2).$

(1)   The proof will be easier by making a diagram like above, and by marking the first two premisses on it.

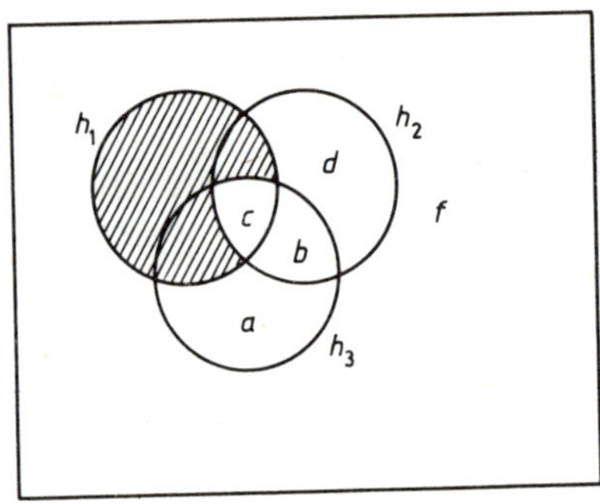

Figure 7

(2) According to $P(h_2 \vee h_3, k) \leq P(\sim(h_2 \vee h_3), k)$,
$$a+b+c+d \leq f.$$

(3) Since $P(h_1, k) > 0$, $c > 0$. Hence, applying
$$a+b+c+d+f = 1, \ 0 < f < 1.$$

(4) From (2) and (3) it follows that
$$b+c+d > 0, \ d+f > 0.$$

(5) Hence
$$(b+c+d)(d+f) > 0.$$

(6) Let us introduce the following extension of (5):
$$(a+b+c+d+f)a + (b+c+d)(d+f) + ad > ad.$$

(7) The left side of the inequality can be transformed into
$$(a+b+c+d)(a+d+f) > ad.$$

(8) Using the equivalences $a+b+c+d = f$ and
$a+d+f = b+c$:
$$f(b+c) > ad, \ bf + cf - ad > 0,$$

(9) which yields, through addition,
$$(a+b+c+d+f)a + bf + cf - ad > a.$$

(10) The left side can be transformed into
$$(a+b+c)(a+f) > a.$$

(11) Hence
$$a+b+c > \frac{a}{a+b},$$

(12) which corresponds to the inequality to be proved:
$$P(h_3, k) > P(h_3, k \ \& \ \sim h_2).$$
Rules $V'_D 6$ to $V_D 9$ are proved similarly.

$V'_D 6$ $\{P(h_1 \supset h_2, k) = 1, \ P(h_3 \supset h_2, k) < 1, \ P(h_1, k) = P(h_3, k)\} \Rightarrow$
$\Rightarrow P(h_1, k \ \& \ h_2) > P(h_3, k \ \& \ h_2)$.

$V'_D 7$ $\{P(h_1 \supset h_2, k) = 1, \ P(h_3 \supset \sim h_2, k) = 1, \ P(h_1, k) > 0,$
$P(h_2, k) < 1\} \Rightarrow P(h_1, k \ \& \ \sim h_2) = 0, \ P(h_3, k \ \& \ \sim h_2) >$
$> P(h_3, k)$.

V'_D8 $\quad \{P(h_1|h_3, k)=1,\ P(h_1 \supset h_2, k)=1,\ P(\sim h_3 \supset h_2, k)=1,$
$P(h_1, k)=P(h_3, k)>0,\ P(h_2, k)<1\} \Rightarrow P(h_1, k\ \&\ h_2)>$
$> P(h_1, k)=P(h_3, k)>P(h_3, k\ \&\ h_2).$

V'_D9 $\quad \{P(h_1|h_3, k)=1,\ P(h_1 \supset h_2, k)=1,\ P(\sim h_3 \supset h_2, k)=1,$
$P(h_1, k)=P(h_3, k)>0,\} \Rightarrow P(h_1, k\ \&\ \sim h_2)=0,$
$P(h_3, k\ \&\ \sim h_2)=1.$

All these inferences are, however, "thin air" until we can provide rules in order to assign a comparative value of probability to every hypothesis.

7 RULES OF VALUATION

This chapter is designed to propose answers to the following questions:

(1)   When can we assign the value 0 to a hypothesis?

(2)   When can we hold the probability of some hypothesis to be 1?

(3)   When can we say that a hypothesis is of the same probability as another?

(4)   When is the probability of a hypothesis at least as high as that of another?

In each case the same background of knowledge is presupposed.

It is possible to furnish some rules on purely logical grounds for the valuation of hypotheses.

$$\text{L1 If } k \Rightarrow p, \qquad \text{then} \quad P(p, k) = 1.$$

$$\text{L2 If } k \Rightarrow \sim p, \qquad \text{then} \quad P(p, k) = 0.$$

$$\text{L3 If } k \Rightarrow p \equiv q, \qquad \text{then} \quad P(p, k) = P(q, k)$$

$$\text{L4 If } k \Rightarrow p \supset q, \qquad \text{then} \quad P(p, k) \leq P(q, k).$$

It is easy to notice that L1 is the consequence of T1—T3, and L2 is that of T4—T6. Similarly, L3 can be grounded on T17, while L4 on T14.

Replacing "if and only if" for "if" in L1 and L2 is permissible, though not obligatory. In this way it is obvious that the above rules are sufficient but not necessary for the valuation of hypotheses. It remains a question whether they are efficient in the comparative valuation of

4*

hypotheses. Scientific cognition in practice tends to indicate that they cannot be considered as adequate answers to the respective questions. It may happen that a hypothesis p or its negation $\sim p$ cannot logically be deduced from the background knowledge k, yet p must be assigned a probability value 1, or 0. Moreover, some hypotheses compared may often have implications or display equivalences which do not follow from the background knowledge, still the hypotheses must be credited with the same logical probability. Therefore it seems to be warranted to add some methodological rules of valuation to the above outlined logical ones.

The synchronical inferences applied to the revision of our hypotheses have premisses whose logical probability is maximal, although they can be considered neither as logical consequences of the background knowledge nor to be logically true. (Cf. V_S1) These premisses are taken to be true with certainty on the grounds that no rational doubt is permitted about them on the given level of cognition. They could just as well be included in the background knowledge, too. If this were done, the hypothesis in question would be a logical consequence of the background knowledge. (Cf. L1) In that case, however, it would be necessary to change the initial background knowledge and the actual synchronic (deductive) process of inference would not become explicit either. To counter these disadvantages, maximal relative truth values are assigned to the part of earlier reliable knowledge that is relevant to the argument, while the rest of it is only regarded as background knowledge. The relevant part is usually a theory or a law from which the theorem to be proved can be logically deduced. These considerations lead to the following methodological rule.

M1   If a hypothesis h is not a consequence of the background knowledge k nor is it logically true, it can only be considered as true with certainty if it is a theory or law that could also be included in the background knowledge.

It has been demonstrated that probabilistic inferences can only be applied to the revision of hypotheses with probabilities higher than minimal. Nevertheless, in the practice of scientific reasoning, a significant part of the hypotheses deemed to be of more than 0 probability on the basis of L2 will have their probability reduced to 0 because of certain defects in their contents. In other words, minimal probability is attributed not only to hypotheses which are self-contradictory (or internally inconsistent), or are totally incompatible with the background knowledge (externally inconsistent), but to those which share certain epistemological defects as well. Such hypotheses fail to meet some elementary requirements of contents among the criteria of being scientific. They are often said to be unscientific, too. Now we are introducing some methodological rules to mark certain indices of hypotheses suspected of being unscientific.

M2   A hypothesis shall not contain any hypostases.

Hypostases, attributing real existence to abstractions, imply Platonism. To use a term of mediaeval philosophy, hypostatization can be called extreme realism as it "adds" fictitious existents to the realm of real existents and thus it duplicates reality. Hypostases are e.g. such existents as "primary matter", "essential form", "entelechy" (in the Aristotelian sense), or the souls of plants, animals, and human beings as substances. It is also hypostatization to suppose that the primary existents of reality are logical or mathematical objects (e.g. differential equations).

M3   A hypothesis shall not contain phenomenalistic
            reductions.

Phenomenalists reduce spheres of reality beyond experience to the reality of experience and the latter to the sense data of experience. This in fact is an application of Berkeley's principle "esse est percipi", through which the existence of a reality independent of the cognitive subject is ultimately rejected as a metaphysical presupposition. This approach is met with in the philosophy of E. Mach, in the so-called Copenhagen interpretation of quantum mechanics, or in behaviourist psychology.

M4 A hypothesis shall be level congruent.

It is a view integrally embodied in the modern scientific world picture that reality has simple as well as complex levels of organization, which significant qualitative differences. There are physico-chemical, biological, psychological, and socio-cultural levels of organization. These levels include areas that are genetically related to each other and constitute a line of development. Such are for example the stages marking the emergence of man from the animal kingdom. In our opinion it is not tenable as a hypothesis to reduce developed levels of organization to less developed ones, or to project higher forms of development onto lower levels. Examples of gross negligence of the principle of level congruence were mechanistic conceptions of man and society, geographical determinism, animistic conceptions of nature, or the myth of thinking machines.

M5 A hypothesis shall be testable, at least in principle.

Testability means that the hypothesis yields logical consequences that can be confronted with the findings of scientific experience. Accordingly, the hypotheses put to tests are very often rather complex statements from which, taking them as premisses, and following the rules of deduction of classical two-valued logic and of mathematics, we can derive nomological hypotheses or individual statements which can be judged either true or false on the grounds of empirical data. In this way, it will be possible to justify or refute a hypothesis. It cannot be set as a demand, however, that justification become actually possible within definite limits of time. Testability in principle is only required. This condition must be met if the prior probability of a hypothesis is to be higher than 0, i.e. failure to meet it is sufficient grounds for us to suspend the hypothesis as being of minimal probability.

Rules M2—M5 are meant to provide for an epistemological screening of logically precluded hypotheses. They are therefore regarded as mandatory value categories prescribed to fulfil the conditions necessary for being scientific.

It is a much more intricate task to attach suitable epistemological considerations to the logical rules L3 and L4. The epistemological criteria needed are such as to make it possible to favour a hypothesis h_1 over another, h_2, against a shared background knowledge k, or in lack of k to judge two hypotheses as equally probable.

Before tackling these rules in detail, we must briefly expose the *principle of ceteris paribus*. This principle, as it were, coordinates the criteria of preference and of indifference, performing in this way a super-methodological function. It does not permit the preference of one hypothesis to another according to a given criterion unless the preferred one is at least as "good" as the other by any other criteria.

This is the narrowest application of the principle of ceteris paribus. We speak of a broader application if a prior order of importance has been set up for the methodological rules. Let us assume that such an order is determined by rules M6—M9 below, in which case the following steps are to be adopted: If, on the basis of M6, h_1 is better than h_2, then h_1 is preferred. If they are equally good, they will be contrasted on the basis of M7, as in the previous step. In case of indifference the comparison is further carried on. This way of application presupposes that the sequence of the methodological rules corresponds to the order of importance of the criteria for comparing rival hypotheses. In this way we can avoid the following difficulty: Assume that, according to M6 and M7, hypothesis h_1 is "better" than h_2, whereas h_2 is better than h_1 according to M8 and M9. If the sequence in application of the methodological rules were arbitrary, neither of the hypotheses could be preferred to the other. The principle of ceteris paribus, however, allows us with certainty to assign a higher logical probability to the hypothesis h_1.

Let us see now the criteria of preference.

M6   If hypothesis h_1 is more general than h_2, then it is ceteris paribus less probable.

Here by generality we mean the range of the spheres of reality that the hypotheses under investigation purport to cover. For example, geometrical optics is less general than physical optics. Similarly, the theory of relativity is of greater generality than classical mechanics.

When comparing hypotheses in mathematical formulation we may examine the invariance and permanence of hypotheses as indices of their generality. If e.g. the mathematical formulation of the hypothesis h_1 is invariant to Lorentz transformations while the mathematical formulation of h_2 is not, then h_1 is more general and therefore less probable than h_2. The principle of performance on the other hand asserts that laws valid for a limited area remain valid for a broader one as well.

M7   If the hypothesis h_1 is deeper than h_2, then it is ceteris paribus less probable.

There are some phenomenological or one-dimensional hypotheses which only grasp the surface of the objects examined, seeking to generalize facts of experience in the form of nomological hypotheses. Other hypotheses go beyond the sphere of experience in an attempt to reveal deeper or so-called intrinsic mechanisms, for only in this way is it possible for us to create an adequate picture of objects structured on several levels. These hypotheses are called structural or model hypotheses. It is a well-known fact that the periodicity of chemical elements was at first thought to be explained by changes in their atomic weights. However, this was quite a superficial explanation since it could not determine a mechanism through which atomic weights could effect the regular recurrence of chemical characteristics. Only quantum theory came to make that possible, according to which the prior probabilities of hypotheses explaining chemical phenomena by quantum theory are ceteris paribus smaller than those of earlier phenomenological explanations. Any increase in depth reduces the chances of a hypothesis to be true.

M8   If the hypothesis h_1 is simpler than h_2, it is ceteris paribus less probable.

A hypothesis is said to be simpler than another if it contains fewer categories and lower order logical and mathematical operations. Consequently the construction of theoretical hypotheses must not include superfluous entities (Occam's razor) or an apparatus more complicated than warranted. As the well-known old saying has it: Simplex sigillum veri. Our above principle contradicts that maxim, though: the simpler a hypothesis the less its relative truth. Still the contradiction is only a partial one since in fact an inverse relation holds between the prior probability and the simplicity of a hypothesis. Our position is not far from Popper's who holds that the simpler a hypothesis is the more falsifiable it is (Popper, 1959, 137—143). A. A. Simon has introduced the limitation to this application of the principle of simplicity that a hypothesis must be more than a mere generalization of data (Simon, 1968, 439—459). We shall nevertheless see that a hypothesis of a lower prior probability, when justified, will be more likely to be accepted as new and true knowledge than one of a high prior probability. The contradiction is thus resolved in the frame of the diachronical approach.

M9   If the hypothesis h_1 is more exact than h_2, then it is ceteris paribus less probable.

If a calculation is required to be exact to one decimal place, it has a greater chance to be correct than if it is required to be exact to two or more places of decimals. In the same way, a proposition expressing a possibility can be true when a statement of fact or necessity is false. Thus the former will be attributed greater probability. Hypotheses that arc logically stricter are less probable. Logical strictness is a special form of exactitude.

Generality, depth, simplicity, and exactitude constitute the so-called *intrinsic values* of hypotheses. Ultimately these values are a measure of the chance of a hypothesis to add new knowledge to thc background knowledge if it has been justified, i.e. a measure of the degree to which the given hypothesis copes with the expectations for a hypothesis to be "good". M6—M9 are optative categories.

Much of our elaboration of these methodological rules has been based on the ideas of M. Bunge (Bunge, 1967, II, 346—356) and J. T. Davies (Davies, 1973, 94—104 and 162—173). Bunge sets up such rules as obligatory or optatively satisfied to ensure the acceptability of a hypothesis. Davies applies similar rules to the evaluation of the degree of confidence of hypotheses and makes an attempt to introduce nominal quantification. Our views are fundamentally different from these ideas since our methodological rules are merely meant to offer some guidelines for the rational pre-valuation of hypotheses. This kind of preliminary evaluation, if the hypotheses are tested through diachronical reasoning, may be more or less modified in keeping with the expansion of background knowledge.

The methodological rules discussed in the foregoing are based on regularities drawn from the history of scientific cognition, regularities which indicate the line, or at least the tendency, along which scientific cognition may proceed. Accordingly, they are of more than a merely descriptive character as they appear as necessities or desirable tendencies deeply rooted in the nature of science and therefore they assume a normative role. We believe that the basic goal of methodological studies into the history of science is to find out how scientific thinking corrects itself from time to time and to draw up better and better cognitive strategies making use of past experiences of both success and failure. One of the least developed among these strategies is the prevaluation of hypotheses. Diachronical reasoning is impossible without suitable rules of valuation.

However, in order to form value judgements of hypotheses it is not enough to derive certain unknown probabilities from probabilistic inferences: it is prerequisite to take account of the cognitive approaches of arguers as well.

8 COGNITIVE APPROACHES

Throughout the elaboration of the thoughts presented in this chapter we made extensive use of the conceptions of K. Mannheim (Mannheim, 1929), K. Lewin (Lewin, 1935, 1—42), L. Festinger (Festinger, 1957), G. W. Allport (Allport, 1954), and M. Maruyama (Maruyama, 1960, 1974). Our basic source of inspiration was at the same time the idea of a dialectical epistemology that would consider its main task to be the description of the growth of knowledge.

Let us start our further investigations with the contention that the long history of scientific cognition not only provides grounds for some methodological rules but also impresses certain cognitive approaches upon cultivators of the sciences. On the one hand these approaches reflect objective features of some cognitive situations and on the other, they idealize those features and guide the researchers accordingly by causing them to develop certain attitudes and so to be prepared for various reasoning situations. We shall henceforward generally assume the arguers' cognitive approaches to be right, i.e. that arguers will not base their reasoning strategies on prejudices. As a consequence, no differences between the cognitive approaches of arguers are permitted. This is of course an idealization far removed from real situations of reasoning, nevertheless it is prerequisite to the rationality of argumentation.

Towards the end of this work we shall also examine some cases displaying a lack of uniformity in the cognitive approaches of reasoners, i.e. when at least one of them misconceives the real nature of the given cognitive situation. This, however, does not always prevent the participants of an argument from taking a uniform point in accepting or rejecting a hypothesis.

There are cognitive approaches that provide *long-term* orientation, and others providing *short-term* orientation.

Among long-term epistemic value orientations let us first of all point to the tendency of *seeking truth*, which means the same as *to avoid errors*. The specific approach basically described as seeking the truth is called *cognitivism*.

Cognitivism at its fullest is found in the field of synchronic argumentation where efforts are made exclusively to seek the truth and to avoid errors.

A scientist seeks not only true but *new (original) knowledge* as well, therefore he tries to *avoid conventionally true knowledge*. If he is dominantly oriented towards new knowledge, he is said to have a *critical* approach. It is quite obvious that, if the critical approach is driven to extremes, it will prevent any reasoning and will be a hindrance to cognition.

The epistemic attitude of a scientist may emerge as the resultant of the above-mentioned approaches. This case is termed *critical cognitivism*. As a basic epistemic value orientation, critical cognitivism can be well identified in the diachronic modes of reasoning. These inferences include hypotheses which have neither maximal initial probabilities nor minimal ones, i.e. which do not force us to choose between the background knowledge and the hypotheses. This cognitive situation arises when the conditions $k \nRightarrow h$ and $k \nRightarrow \sim h$ are fulfilled.

Each of these types of value orientation can be related to corresponding *strategies of reasoning*. (Cf. 13) In order to sufficiently characterize these strategies we must examine the approaches or attitudes that reflect the specific features of concrete reasoning situations and, as close value orientations, immediately determine the (short term) direction of the reasoning. Among them the first to be discussed will be the attitudes related to the emergence of cognitive change.

The history of science attests to the occurrence of cognitive situations which essentially lead to the expansion of human knowledge. This period of change in science is now customarily associated, following T. Kuhn, with the *cumulative* approach (Kuhn, 1962). The rise of some important scientific discovery is often followed by a peaceful period of

60

development which is devoted to the task of elaborating the new theory and unfurling the new thoughts inherent in it. It may as well happen that the conception of cumulative cognition is, in a given cognitive situation, quite warranted since the actual nature of the change in cognition is adequately reflected in it. The cumulative phase of cognition embraces both the background knowledge representing past knowledge and the new knowledge embodied in the hypothesis to be tested.

When a branch of science is at its stage of "peaceful" development and if this is properly recognised by the arguers, their basic cognitive goal will appear to be maintaining the continuity of cognition and expanding knowledge accordingly. This epistemic attitude is called the *orthodox approach.* (Where the concept of orthodoxy is meant to imply no sense of contempt but to designate an approach corresponding to the conservative phase of scientific cognition.)

It is easily conceivable at the same time that the participants of the reasoning do not adequately recognise the character of the given period of scientific cognition, entertaining the belief that what is needed is radical change and the challenging of tradition. This attitude is termed the *relativistic approach.*

There are then, to be sure, cognitive situations which are really revolutionary and are basically characterized by a challenge to earlier knowledge. However, the moment of interruption must not imply that old and new knowledge are wholly incompatible. The approach which is essentially adequate in reflecting the revolutionary character of scientific cognition is called the *heterodox attitude.*

Not infrequently, there are scientists who remain with mistaken conceptions as to the given phase of cognition: when the breach with earlier views has in fact been long overdue, they are still ardently working towards the conservation of tradition. These scientists are aptly described by M. Planck, who explains that a scientific truth is not usually established because its adversaries are eventually persuaded and converted, but rather, the adversaries gradually die and a new generation is brought up in the truth from the very beginning (Planck, 1948, 22). The attitude so characterised will be labelled *epistemic dogmatism.*

Just as in politics *the principle of the concrete analysis of the concrete situation* is of basic importance for elaborating the best strategy and tactics, safeguarding against doctrinarianism, so is the concrete and unbiased investigation of the actual cognitive situation in scientific research an indispensable precondition to the dialectical approach. Concreteness, however, presupposes a totalistic approach, i.e. regarding the given cognitive situation as part of the process of the development of cognition. We do not believe it is necessarily a sign of consistently representing the dialectical approach if a scientist presses the idea of development under all circumstances irrespective of the objective nature of the actual situation. The dialectical approach does not imply unrestrained claims to novelty but the adoption of strategies corresponding to the actual movement of reality. In this way, the orthodox view is not to be identified with dogmatism nor the heterodox attitude with relativism since they reflect some real tendencies of the development of science. Both dogmatism and relativism are, on the other hand, metaphysical approaches because they replace the concrete analysis of the concrete situation with the abstraction and over-generalization of one single tendency.

The way to search for new knowledge then depends on the cognitive situation in which scientific reasoning takes place and in fact facilitates the progress of cognition only if the participants of the argumentation share a consistently dialectical approach i.e. their epistemic attitude is situation dependent.

There are cognitive situations where the aspect of continuity is dominant (orthodoxy) and others in which the aspect of discontinuity is decisive (heterodoxy). These situations comprise the two extreme kinds of cognitive situations. Between these two opposites there is a scale of transitory situations which display a more or less perfect balance of continuity and discontinuity.

Correspondingly, a continuum of the quest for new and true knowledge can be supposed to connect the orthodox and the heterodox attitudes. A cognitive situation may be said to be wholly favourable either to the orthodox attitude, or to the heterodox approach, or just more favourable to the one than to the other, but it may also require a *balance* between the two attitudes. In the last case we speak of *neology*.

Not only are the value judgements of an arguer thoroughly influenced by the fact whether the development of knowledge is dominated by continuity or discontinuity but by the significance he attributes to the different stages of the process of cognition, too. Some scientists hold that the expansion of earlier knowledge can be expected almost exclusively of new empirical knowledge only, so they work out a cognitive strategy that is declared to aim at closeness to experience. This epistemic attitude is said to be *radically empirical.*

Others expect the growth of knowledge to ensue mainly from more and more adequate reconstructions in thought of the objects under investigation, reconstructions which are only possible by theoretical means. This approach is labelled *radically theoretical.*

Yet another approach is of course also encountered, in which the above mentioned radical conceptions are levelled, as it were, in a dialectical synthesis. This is simply called an *empirical-theoretical* approach.

After such considerations, we surmise that a radically empirical approach does not permit the acceptance of hypotheses unless they are certain, i.e. they have received maximum empirical justification. This attitude is one of minimizing the role of theory in an effort to confine scientific research to the limits of gathering and arranging mere data. However, this approach cannot be considered to be one-sided altogether, as there really are situations recurrent in the history of science in which the task of obtaining reliable data and furnishing empirical generalizations of a relatively low order is emergent. This period is normally called a *prescientific or descriptive stage of development.*

At a higher stage of its development, the main concern of science becomes the construction of theories i.e. of systems functioning as wholes made up of more or less isolated generalizations. What in fact corresponds to an actual research situation is the radically theoretical approach. A case of the joint occurrence of the heterodox and the theoretical attitudes is called a *scientific revolution.*

The rise to the theoretical level in science is usually followed by a phase of development which unfolds the developmental possibilities accumulated during the earlier periods and evolves the basic core of theory that has already been formed but not yet expounded. Corresponding to this phase is the empirical-theoretical attitude which, if associated with the above presented neological approach, allows us to speak of a so-called *gradual or normal development* of science.

As it is quite frequently the case, research workers often form in themselves a false image of the given period of the development of knowledge. Some strive to obtain the most possible empirical data as an end in itself, with no regard to the nature of the research situation, going as far as to cause a regular "information boom". This disposition is normally called an *empiricistic* or *anti-theoretical* approach.

In a similar fashion, one is not always warranted to apply a radically theoretical approach since a branch of science may often prove to be immature for being theorized. Despite this fact, there are scientists who make efforts to press the idea of theorizing even in such cases. They may be rated as *speculative* or *idle theorists*.

Finally, it is not uncommon that a scientist decides to follow the "media via" indepently of the nature of the situation and wishes to contribute to the promotion of the theory by drawing new logical conclusions from the principles laid down previously. His attitude is known as *scientific epigonism*.

As it has been stressed, for the time being we suppose arguers to have properly apprehended which stage of development of scientific cognition the given reasoning situation belongs to.

Before we can carry on our investigation we must face an objection that may be raised to the foregoing. When analysing the comparability of hypotheses we pointed out that the homogeneity of the background knowledge requires any new knowledge to be connected as a conjunctive constituent to the old one, which means we know more about a limited area of reality. The present analysis, however, allows for a heterodox approach in which the necessity of radical change is expressed. The problem is the apparent incompatibility of the cumulativity required by the homogeneity of the background knowledge with the non-cumulativity associated with the heterodox approach.

This problem can be resolved if we consider that the object of the above outlined cognitive approaches is not the nature of the background knowledge but the way knowledge is extended and the character of hypotheses as potential knowledge. An attitude is always oriented towards the future and not the past. It is a different matter that a well justified hypothesis formed through the heterodox approach cannot be connected as a conjunctive component to the prior background knowledge: it functions as new background knowledge to which both earlier and later knowledge must be compared.

The attitudes concerning the nature of change in knowledge and the given stage of its development do not describe the specific character of reasoning situations with sufficient precision. Another parameter is to be taken into account: *the minimal degree or value norm of acceptability.*

In some cognitive situations the value norm of acceptability is low. This means that scientists take great risks of error in passing their value judgements, i.e. they take bold decisions. If they set a high value norm of acceptability, they adopt a *cautious* attitude. Finally there are attitudes taking *moderate* risks of error, reconciling the bold and the cautious approach.

However, it remains a question when it is rational to take one of these risks. The answer depends on *the importance attributed to the simple reproduction of the background knowledge* in the given situation of reasoning. If the mere reproduction of background knowledge counts as a great scientific achievement, the argumentation is said to be cautious, while if the value of mere reproduction is rated low, it implies a bold reasoning behaviour.

It may be noted that the risk of error is sometimes miscalculated. Such mistaken behaviour is called *rash, timid,* or *compromising.* These, however, will be excluded from further study.

The epistemic approaches described earlier will also be integrated in the field of reasoning. The simplest among them to handle are the parameters expressing the divergence from reality and the risk of error. Let α signify the former, and λ the latter. For simplicity's sake, let us assume that

$$0 \leq \alpha \leq 1$$

$$0 \leq \lambda \leq 1.$$

In accordance with our previous suggestions, the different approaches can be defined as follows:

An empirical approach implies $\alpha=0$.

A theoretical approach implies $\alpha=1$.

An empirical-theoretical approach implies $0<\alpha<1$.

A cautious approach implies $\lambda=0$.

A bold approach implies $\lambda=1$.

A moderate approach implies $0<\lambda<1$.

The approaches concerning the nature of change in cognition cannot be characterized by a single parameter. The case is that in different cognitive situations, the notion of novelty or originality is attributed substantially different senses. There must be suitable measures found in order to evaluate the novelty of hypotheses. For this purpose, we must introduce the notion of semantic information.

9 THE INFORMATION CONTENT
OF A HYPOTHESIS

An important role in judging the acceptability of hypotheses is played by the information contents they convey: the less the probability of a given hypothesis, the more its information content. This should be natural, since the less the background knowledge allows us to expect the truth of a hypothesis, the more the information carried by our learning of it. Correspondingly, the measurement of information content is only possible by a function which changes inversely with probability (the relative truth value).

Among functions measuring the information content, the following are normally in use:

$$\text{d1} \qquad S(p, k) = -\ln P(p, k)$$

$$\text{d2} \qquad C_1(p, k) = P(\sim p, k) = 1 - P(p, k).$$

The function S defined by d1 is interpreted like this: $S(p, k)$ expresses the *surprise value* of the hypothesis p in relation to the background knowledge k (Carnap–Bar-Hillel, 1964, 221—274; Hintikka, 1968, 311—331). However, the measurement of information content in the philosophy of science is more customary by the function C_1 than by S. C_1 is called the *degree of novelty* or *entire new content* (Hintikka, 1968, 313—317). The connection between the above mentioned kinds of information is thus expressed:

$$S(p, k) = \ln \frac{1}{1 - C_1(p, k)} = -\ln P(p, k).$$

The function C_1 can be divided into two constituents by the use of T8 as in this equality:

$$C_1(p, k) = C_2(p, k \,\&\, q) + C_3(p, k \,\&\, q).$$

The functions C_2 and C_3 are defined as follows

$$\text{d3} \qquad C_2(p, k \,\&\, q) = C_2(p \,\&\, q, k) - C_1(q, k)$$

$$\text{d4} \qquad C_3(p, k \,\&\, q) = C_1(p, k) - C_2(p, k \,\&\, q).$$

It is easily recognized that

$$C_2(p, k \,\&\, q) = P(q, k) - P(p \,\&\, q, k) = 1 - P(p \vee \sim q, k) =$$

$$= 1 - P(q \supset p, k) = C_1(q \supset p, k)$$

$$C_3(p, k \,\&\, q) = 1 - P(p \vee q, k) = C_1(p \vee q, k).$$

$C_2(p, k \,\&\, q)$ expresses the information that the hypothesis p can provide on the basis of k without the additional knowledge q. Therefore, it is usually called *incremental content*. On the other hand, $C_3(p, k \,\&\, q)$ is termed *common content* because it expresses the new content provided by p and q together. To make the relation between the above-mentioned kinds of information clear, let us suppose that the additional knowledge q, itself a hypothesis at the outset, becomes justified. Then the part becoming useful information of the total new content $C_1(p, k)$ is the one provided by k and p together, while only that fragment can be utilized later which is furnished by p alone, i.e. $C_2(p, k \,\&\, q)$. Let the relation e.g. $k \Rightarrow p \supset q$ hold. Then $P(p \supset q, k) = 1$ and the distribution of the total new content $C_1(p, k)$ is

$$C_2(p, k \,\&\, q) = P(q, k) - P(p \,\&\, q, k) = P(q, k) - P(p, k)$$

$$C_3(p, k \,\&\, q) = 1 - P(q, k).$$

The refutation of p results precisely the opposite distribution of the information:

$$C_2(p, k \,\&\, \sim q) = 1 - P(q, k) + P(p \,\&\, \sim q, k) = 1 - P(q, k)$$

$$C_3(p, k \,\&\, \sim q) = 1 - P(p, k) - 1 + P(q, k) + P(p \,\&\, \sim q, k) =$$

$$= P(q, k) - P(p, k).$$

Incremental information is interpreted as an index expressing the continuity in the growth of cognitive content. It is a suitable device for the successive evaluation of the content increasing effect of each hypothesis alone. Common content may then be considered a measure of discontinuity because it does not separately display the content growing effect of the hypotheses examined as against the background knowledge. The effect emerges as a whole, "in one unit".

It has already been explained that the epistemic approach which is exclusively interested in the continuity of cognition is qualified as orthodox. An orthodox arguer only judges the informative value of hypotheses on the grounds of the incremental content, with no regard to the common content. There are cognitive situations where this approach is justified. In these cases the total new content conveyed by p is entirely coming from p alone, p and q having no common content. From this, however, it does not follow that q is not informative to k, i.e. $C_1(q, k) = 0$, since if $C_3(p, k \& q) = 0$, i.e. $P(p \vee q, k) = 1$, then $P(q, k)$ does not necessarily take on the value 1. But if $C_3(p, k \& q) = 0$ then $C_3(q, k \& p) = 0$, that is, if p has no content common with q as compared to k, then q has no content common with p in the same context.

In a similar fashion, the heterodox approach only recognizes the discontinuity of cognition as important. This is justified if none of the hypotheses examined possesses a sole effect to increase contents, i.e. $C_2(p, k \& q) = 0$ and/or $C_2(q, k \& p) = 0$, and thus $P(p \vee \sim q, k) = 1$ and/or $P(\sim p \vee q, k) = 1$. In these cases $C_3(p, k \& q) = C_1(p, k)$, or $C_3(q, k \& p) = C_1(q, k)$.

The neologistic approach takes both continuity and discontinuity into consideration. A way to do this is simply to regard the entire new content as the sum of incremental and common content, which means the roles of the different constituents are not weighted. But another way is also possible, in which the incremental and the common content are weighted according to the nature of the cognitive situation. This issue will be clarified later. Now we shall turn to the notion of epistemic utility and its measurement.

10 EPISTEMIC UTILITY

There are several kinds of utility known. We are now concerned with the specific kind that is relevant to the growth of human knowledge. New scientific information is useful in the sense that it tends to satisfy the general human need for knowledge manifested in the form of problems. Somewhat imprecisely, this kind of utility is called *epistemic utility*.

Let us start from the intuitively acceptable thesis that the acceptance of a true statement constitutes a *gain* while that of a *false* statement is a *loss* from the point of view of cognition. Since logical truths can be recognized independently of empirical evidence just by knowing the relevant logical theory, their acceptance is less useful than that of other truths. So can logical falsities (contradictions) be recognized theoretically, which makes their acceptance less harmful than that of other falsities, our mistake being eliminable by purely theoretical means. Obviously, the richer the contents of a true statement the greater its epistemic use and, similarly, the poorer the contents, the more harmful the acceptance of a false statement is. On these intuitive grounds we can introduce functions to measure *epistemic utility*. Let $C(h, k)$ be the measure of the degree of novelty (of content) of the statement h as compared to the background knowledge k. (For C we can substitute e.g. the function C_1 introduced in the preceding section.)

Consider that in relation to the given background knowledge k, all logical consequences of k (and thus k itself) are equivalent to logical truths, while statements incompatible with k (and thus $\sim k$) amount to logical falsities. Therefore, in relation to k, logical truths can be replaced by k and contradictions by $\sim k$. This will give rise to a considerably more realistic interpretation than is obtained by the discussion, as by

I. Levi of the positive or negative epistemic utility due to the (proper) recognition of the truth of tautologies or to the failure to recognize the falsity of contradictions (Levi, 1967, 77—82). If the background knowledge k really is true and is accepted as being so, the positive epistemic use (gain) arising from the cognitive act may be seen as resulting from the repetition of prior knowledge. It is not impossible, however, that the background knowledge is not accepted as true even though it is, which means that its negation is accepted as such though it is false, but considering it as bona fide true. It is clear that the latter kind of cognitive act implies negative epistemic use (loss).

Let $U(h, 1, k)$ signify the epistemic gain arising from the acceptance of the *true h* as against the knowledge implied by k, and let $U(h, 0, k)$ stand for the negative epistemic utility (loss) resulting from the acceptance of the *false* statement h in relation to the background knowledge k. They can be applied to express the above considerations in mathematical form like this:

(1)
$$\begin{cases} U(h, 1, k) - U(k, 1, k) = \lambda[C(h, k) - C(k, k)] \\ U(h, 0, k) - U(\sim k, 0, k) = \lambda[C(h, k) - C(\sim k, k)] . \end{cases}$$

λ is a proportion factor expressing the degree of boldness, interpreted in the interval $[0, 1]$. The arguments '1' and '0' signify the truth or the falsity of the accepted statement, this time in relation to a definite state of affairs in reality instead of the background knowledge k. Correspondingly, 0 and 1 here are not relative (probabilistic) truth values but the ones presupposed by classical two-valued logic.

As for the gain of accepting logical truths and the loss due to accepting contradictions, we can conclude that

(2)
$$U(k, 1, k) \leq U(h, 1, k)$$

(3)
$$U(\sim k, 0, k) \geq U(h, 0, k) .$$

Thus $U(k, 1, k)$ is the lower limit of positive epistemic utility, while $U(\sim k, 0, k)$ is the upper limit of negative epistemic utility (epistemic loss). This requires

(4)
$$U(\sim k, 0, k) \leq U(k, 1, k)$$

to be fulfilled. If the two values do not coincide, the part of the utility scale between them becomes "inactive": neither $U(h, 1, k)$ nor $U(h, 0, k)$ falls within this range. (However, this may not impose the prescription of the equality $U(k, 1, k) = U(\sim k, 0, k)$ to obtain.)

Now it is possible to interpret the equalities (1). The result of the subtraction $U(h, 1, k) - U(k, 1, k)$ is called the incremental gain resulting from the acceptance of the true hypothesis h in the cognitive situation identified with the background knowledge k. The difference of $U(h, 0, k) - U(\sim k, 0, k)$ is called the incremental loss caused by the acceptance of the false hypothesis h in the same situation. The greater the proportion factor of boldness λ and the more informative h, the greater the incremental epistemic gain. The incremental epistemic loss is great if the value of λ is high and h is less informative. In other words: if we undertake a great risk of error, the application of a hypothesis of a low relative truth value, provided it is true and accepted, will yield a great incremental epistemic gain while in the case of a hypothesis of a high relative truth value, supposing it is accepted though false, we must expect a great epistemic loss.

The following is an examination of the U function in general (without furnishing a concrete C). It is a self-evident convention that epistemic loss is not to be expressed by a positive numerical value nor (epistemic) gain by a negative number. This follows with certainty from (2), (3), (4) if the equality

$$(5) \qquad\qquad U(\sim k, 0, k) = 0$$

is accepted.

However we choose the function C, we can suppose that its values will not exceed the closed numerical interval of 0 and 1, and that

$$C(k, k) \leq C(h, k) \leq C(\sim k, k).$$

These conditions are satisfied for the function C_1. If it is stipulated that

$$(6) \qquad\qquad U(k, 1, k) = \sigma = 1 - \lambda \qquad (0 < \lambda \leq 1)$$

then the values of the function U are confined to the closed numerical interval ranging from -1 to 1. These conventions lead to the following scale of epistemic utility:

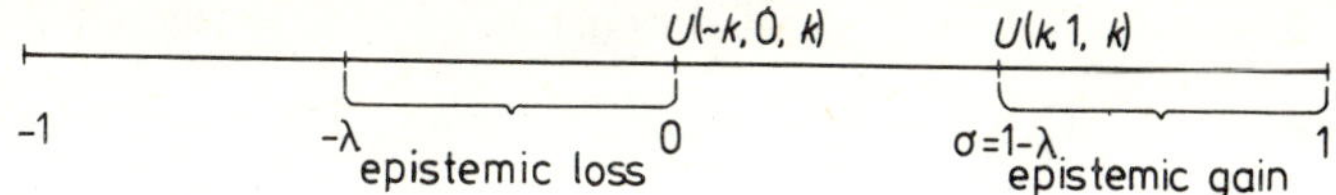

Figure 8

It is clear from Figure 8 that no value of the function U can fall in either the open numerical interval 0 to $1-\lambda$ or in the one -1 to $-\lambda$.

In choosing λ, two extreme cases may occur:

(a) $\lambda=0$. Then for every h, $U(h, 1, k)=U(k, 1, k)=1$ and $U(h, 0, k)= =U(\sim k, 0, k)=0$. This valuation assigns the gain 1 equally to the acceptance of every truth, and the gain 0 to the acceptance of every falsity.

(b) $\lambda=1$. Then $U(\sim k, 0, k)=U(k, 1, k)=0$ (no numerical interval is missing from the scale of utility). This valuation gives maximum weight to the value $C(h, k)$ i.e. the degree of the novelty of h in computing the epistemic gain (or loss).

It is clearly seen that *the choice of the value of λ is a measure of the risk of error taken by accepting new content.*

With regard to the conventions (5) and (6), the equations under (1) can be rewritten as

(7)
$$\begin{cases} U(h, 1, k)=\lambda[C(h, k)-C(k, k)]+\sigma \\ U(h, 0, k)=\lambda[C(h, k)-C(\sim k, k)] . \end{cases}$$

Substitute the function C_1 for the hitherto undefined function C. The corresponding functions U and E are signified by U_1 and E_1. From the definitions (7) and (8),

$$U_1(h, 1, k)=\lambda[C_1(h, k)-C(k, k)]+\sigma=$$
$$=\lambda[1-P(h, k)]+1-\lambda=1-\lambda P(h, k) .$$

Considering that $C_1(k, k)=0$ and $\sigma=1-\lambda$,

$$U(h, 0, k)=\lambda[C_1(h, k)-C_1(\sim k, k]=-\lambda[1-C_1(h, k)]=$$
$$=-\lambda P(h, k) .$$

73

Substitute k for h in the first equality of (7) and k in the second. We get

(8) $\qquad\qquad U(k, 0, k) = -\lambda \quad \text{and} \quad U(\sim k, 1, k) = 1.$

These numbers are the limit terms of our scale of utility. Negative values fall in the interval $[-\lambda, 0]$ and the positive ones in $[\sigma, 1]$. In the case of $\lambda = 0$, the two intervals shrink to the single points 0 and 1, while in the case of $\lambda = 1$, the two together cover the interval $[-1, 1]$. The equalities (8) themselves imply that, *if k is false*, the acceptance of k constitutes a maximal epistemic loss while its rejection (i.e. the acceptance of $\sim k$) is a maximal epistemic gain—of course not in some absolute sense but *in a cognitive situation characterized by k*. (E.g. in relation to the Ptolemaic world picture as background knowledge, its rejection yields a maximal epistemic gain since it is a false picture.)

The full scale of epistemic utility appears like this:

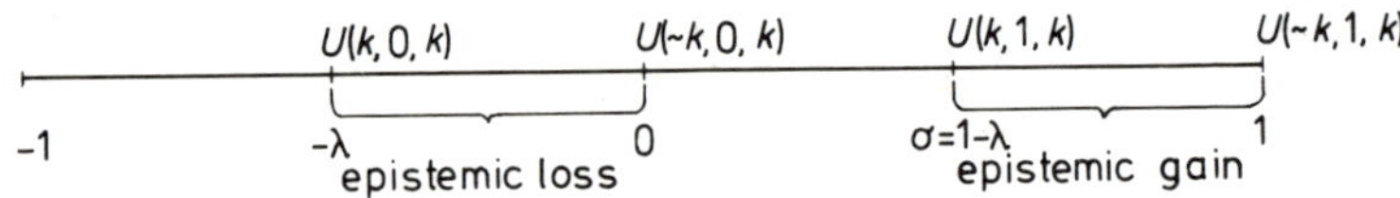

Figure 9

Until now we have only mentioned the epistemic gain arising from acceptance. However, the acceptance of $\sim h$ or $\sim k$ corresponds to the rejection of h or k. Let U^+ stand for the epistemic gains coming from acceptance and U^- for those stemming from rejection. These equalities will result:

$$\begin{cases} U^+(h, 1, k) = U^-(\sim h, 0, k) \\ U^+(h, 0, k) = U^-(\sim h, 1, k). \end{cases}$$

It will not be difficult to define $U^-(h, 1, k)$ and $U^-(h, 0, k)$ either:

(9) $\qquad U^-(h, 1, k) - U^-(k, 1, k) \ = \lambda(C(\sim h, k) - C(\sim k, k))$

$\qquad\qquad U^-(h, 0, k) - U^-(\sim k, 0, k) = \lambda(C(\sim k, k) - C(k, k)).$

74

It is assumed that

$$(10) \qquad U^-(k, 1, k) \; = U^+(\sim k, 0, k)$$
$$U^-(\sim k, 0, k) = U^+(k, 1, k).$$

From (9) and (10), a short calculation leads to the values

$$U^-(h, 1, k) = -\lambda P(\sim h, k)$$
$$U^-(h, 0, k) = 1 - \lambda P(\sim h, k).$$

In virtue of the above, the equalities

$$U^-(h, 1, k) = U^+(\sim h, 0, k)$$
$$U^-(h, 0, k) = U^+(\sim h, 1, k)$$

are easily seen to obtain.

As a result, we can see that the functions U^- can be reduced to the functions U^+ through negation and the exchange of the classical truth values. Therefore below we shall only be concerned with U^+ functions, which will render the sign '+' omissible.

Let us turn our attention to the change in the background knowledge, the shift from k to $k \& p$. Consider that, with regard to p, $C_1(h, k)$ can be divided into the incremental information $C_2(h, k \& p)$ and the conveyed information $C_3(h, k \& p)$. Now we define the functions U_2 and U_3, the former expressing the incremental gain in h compared to p and the latter representing the gain in h on the basis of its useful information content. In the definitions (7), substitute "$k \& p$" for k, and C_2 and C_3 for C:

$$U_2(h, 1, k \& p) = \lambda[C_2(h, k \& p) - C_2(k, k \& p)] + \sigma = \lambda[P(p, k) -$$
$$- P(h \& p, k) - (P(p, k) - P(k \& p, k))] + 1 - \lambda =$$
$$= 1 - \lambda[P(h \& p, k) + P(\sim p, k)] = 1 - \lambda P(p \supset h, k)$$

(in virtue of $P(p, k) - P(k \& p, k) = 0$).

$$U_2(h, 0, k \,\&\, p) = \lambda[C_2(h, k \,\&\, p) - C_2(\sim k, k \,\&\, p)] =$$
$$= \lambda[P(p, k) - P(h \,\&\, p, k) - (P(p, k) -$$
$$- P(\sim k \,\&\, p, k))] = -\lambda P(h \,\&\, p, k)$$
$$(P(\sim k \,\&\, p, k) = 0).$$
$$U_3(h, 1, k \,\&\, p) = \lambda[C_3(h, k \,\&\, p) - C_3(k, k \,\&\, p)] + \sigma =$$
$$= \lambda[1 - P(h \vee p, k) - 1 + P(k \vee p, k) + 1 - \lambda] =$$
$$= 1 - \lambda P(h \vee p, k)$$
$$(P(k \vee p, k) = 1).$$
$$U_3(h, 0, k\&p) = \lambda[C_3(h, k\&p) - C_3(\sim k, k\&p)] =$$
$$= \lambda[1 - P(h \vee p, k) - 1 + P(\sim k \vee p, k)] =$$
$$= \lambda[P(p, k) - P(h \vee p, k)].$$

Epistemic gains and losses are very significant for judging the acceptability of hypotheses. They must be considered together, setting a "balance of gains and losses" for every hypothesis under examination. This is best accomplished by defining the expected epistemic utility.

11 EXPECTED EPISTEMIC UTILITY

On the analogy of the way the concept of expected value is interpreted in probability theory, here we introduce the expected epistemic utility of a hypothesis h in relation to the background knowledge k. Let $E(h, k)$ stand for the definiendum and the definition is

$$(11) \qquad E(h, k) = U(h, 1, k)P(h, k) + U(h, 0, k)P(\sim h, k).$$

The expected epistemic utility of h is then made up of the epistemic gain supplied by the truth of h if it is true, plus the epistemic loss implied by the acceptance of h if it is false, weighted by the probability of the truth or falsity of the hypothesis in question on the basis of the background knowledge k.

It will soon become apparent that the above defined function E will have an important role in the acceptance of hypotheses.

According to definition (9), E depends on U, and U depends on the choice of C. By substituting the values of $U_1(h, 1, k)$ and $U_1(h, 0, k)$, compute the function $E_1(h, k)$:

$$(12) \quad E_1(h, k) = U_1(h, 1, k)P(h, k) + U_1(h, 0, k)P(\sim h, k) =$$

$$= [1 - \lambda P(h, k)]P(h, k) - \lambda P(h, k)[1 - P(h, k)] =$$

$$= P(h, k) - \lambda P(h, k) = P(h, k)(1 - \lambda).$$

The result is identical with the one obtained by I. Levi (Levi, 1967, 83).

The expected epistemic utility of the hypothesis $\sim h$ is computed in a similar way:

$$(13) \qquad E_1(\sim h, k) = P(\sim h, k)(1 - \lambda).$$

In virtue of (14) and (11),

$$E_1(h, k) + E_1(\sim h, k) = 1 - \lambda .$$

Hence

$$E_1(h \vee \sim h, k) = 1 - \lambda .$$

Substitute $k \& p$ for k, and U_2 and U_3 for U in definition (11). Let E_2 and E_3 stand for the functions obtained. These functions express the expected epistemic utility against the background knowledge $k \& p$. More precisely, they represent the epistemic gain secured by revising the hypothesis h on the basis of the additional knowledge p. Consequently, E_2 and E_3 display a diachronic approach while E_1 is a projection of a synchronic one.

$$(14) \quad E_2(h, k \& p) = U_2(h, 1, k \& p)P(h, k \& p) + U_2(h, 0, k \& p) \cdot$$

$$\cdot P(\sim h, k \& p) = [1 - \lambda P(p \& h, k)] \cdot$$

$$\cdot P(h, k \& p) - \lambda P(h \& p, k) \cdot$$

$$\cdot [1 - P(h, k \& p)] = P(h, k \& p)(1 - \lambda) .$$

$$(15) \quad E_3(h, k \& p) = U_3(h, 1, k \& p) \cdot P(h, k \& p) + U_3(h, 0, k \& p) \cdot$$

$$\cdot P(\sim h, k \& p) = [1 - \lambda P(h \vee p, k)] \cdot$$

$$\cdot P(h, k \& p) + \lambda[P(p, k) - P(h \vee p, k)] \cdot$$

$$\cdot [1 - P(h, k \& p)] = P(h, k \& p) - \lambda P(h, k) .$$

However, there is nothing to prevent the substitution of $k \& p$ for k in (11), when the epistemic gain offered by h is considered in a cognitive situation where the statement p has been added to the background knowledge k. After a short calculation, we get

$$(16) \qquad E_1(h, k \& p) = P(h, k \& p)(1 - \lambda) .$$

Having compared (14) and (16):

$$(17) \qquad E_1(h, k \& p) = E_2(h, k \& p) .$$

This is somewhat surprising since we might have expected intuitively that the expected epistemic gain computed on the basis of the whole value of novelty would be greater than the one computed on the basis of the incremental content. We must not forget, however, that (16) takes account of the whole value of novelty possessed by h compared to $k \& p$, while (14) reflects the incremental content conveyed by h based on the proof of p. Thus $C_2(h, k \& p)$ is a part of $C_1(h, k)$, not of $C_1(h, k \& p)$. Another reason why it is not impossible for the equality to hold is that $E_1(h, k \& p)$ and $E_2(h, k \& p)$ express epistemic utilities that are expected, i.e., weighted with probabilities, and which may be equal even if $U_1(h, x, k \& p) \neq U_2(h, x, k \& p)$ (where $x = 1$ or 0) since the weighting may balance the differences.

Some Finnish logicians apply a specific "mixed" formula to compute expected epistemic gains (Hilpinen, 1968, 105—107) which we transcribe like this:

$$(18) \qquad E^x(h, k \& p) = U_1(h, 1, k)P(h, k \& p) +$$

$$+ U_1(h, 0, k)P(\sim h, k \& p) .$$

Hence appropriate substitution and a short calculation results in

$$(19) \quad E^x(h, k \& p) = [1 - \lambda P(h, k)]P(h, k \& p) - \lambda P(h, k) \cdot$$

$$\cdot [1 - P(h, k \& p)] = P(h, k \& p) - \lambda P(h, k) .$$

In this case

$$(20) \qquad E^x(h, k \& p) = E_3(h, k \& p) .$$

Like in (17), weighting with probabilities makes equality possible here too, even though $U_1(h, x, k) \neq U_3(h, x, k \& p)$. It makes this result problematic, however, that different backgrounds of knowledge are referred to in (18): k in the case of computing epistemic utility, and $k \& p$ in weighting with probabilities.

Now, let us try to interpret the functions E introduced above. We have mentioned in passing that $E_1(h, k)$ is employed when the probability of the hypothesis h, the object of reasoning, is stated through synchronic inference (V_S, V'_S). The same can be said of $E_1(h, k \& p)$, with the difference that here a later cognitive situation is taken into account: when the background knowledge k has been extended by the statement p.

$E_2(h, k \,\&\, p)$ and $E_3(h, k \,\&\, p)$ are in turn interpreted as outcomes of the diachronic approach. Thus $E_2(h, k \,\&\, p)$ is taken to be a manifestation of *orthodoxy*, and $E_3(h, k \,\&\, p)$ of *heterodoxy*. Grounds for this view are provided by the fact that $E_2(h, k \,\&\, p)$ has been computed on the basis of $C_2(h, k \,\&\, p)$, the incremental information expressing the continuity of cognition, while $C_3(h, k \,\&\, p)$ representing the discontinuity of cognition has been the basis for $E_3(h, k \,\&\, p)$. The fact that these premisses will yield true conclusions in the forthcoming will further justify the above interpretations.

However, the functions E examined here are not suitable for the representation of neology in case of a diachronic approach. Therefore we suggest to introduce another function $E_*(h, k \,\&\, p)$, defined as follows:

$$(21) \quad E_*(h, k \,\&\, p) = (1-\alpha)E_2(h, k \,\&\, p) + \alpha E_3(h, k \,\&\, p) =$$

$$= (1-\alpha)(1-\lambda)P(h, k \,\&\, p) +$$

$$+ [P(h, k \,\&\, p) - \lambda P(h, k)] \,.$$

Remember that α stands for the degree of the theoretical attitude of the cognitive subject. If $\alpha = 1$, then

$$(22) \qquad\qquad E_*(h, k \,\&\, p) = E_3(h, k \,\&\, p) ,$$

and if $\alpha = 0$, then

$$(23) \qquad\qquad E_*(h, k \,\&\, p) = E_2(h, k \,\&\, p) \,.$$

Since $E_2(h, k \,\&\, p) = E_1(h, k \,\&\, p)$, in orthodox and empiricist attitudes the diachronic approach turns into a synchronic one which has the background knowledge $k \,\&\, p$.

It is further observed that, when $k \equiv k \,\&\, p$, the values of all the functions E are $P(h, k)(1-\lambda)$, i.e. $E_1(h, k)$. Consequently, it is indifferent to the synchronic approach whether the arguers' attitudes are orthodox or heterodox, empiricistic or theoretical. But later it will be seen that the expected epistemic gains provided by the functions $E_1(h, k)$ as well as $E_1(h, k \,\&\, p)$ and $E_2(h, k \,\&\, p)$ are greater than 0 only if $\lambda \neq 1$. As a result, it seems quite obvious that the synchronic approach and, among the diachronic ones, the orthodox approach cannot be bold.

We can further comment that the function E_* includes as border cases both the functions E_2 and E_3 representing the diachronic approach and the function E_1 corresponding to the synchronic one.

Until now, it has not been shown how to apply the functions E to the valuation of the acceptability of hypotheses. To this end, rules of acceptance are to be furnished. Let us assume that the epistemic attitudes of all the participants of an argument are identical, i.e., they all judge the character of the argument as a reasoning situation in a similar way. This is said to be a case of cognitive consonance.

12 RULES OF ACCEPTANCE IN CASE
OF COGNITIVE CONSONANCE

Consider the following situation of reasoning, in which the hypotheses h_1 and h_2 are given as possible answers to a scientific problem. The hypotheses are valuated according to their expected epistemic utility that is, their acceptability will be determined by the expected epistemic gain they carry. Our decision is only considered as rational if the following principles are observed:

(I) A hypothesis is to be accepted only if its expected epistemic gain is at least as great as that of its rival and greater than the norm of acceptability, or its posterior relative truth value is maximal.

(II) Do not suspend the acceptance or rejection of a hypothesis unless its expected epistemic gain equals the norm of acceptability but its posterior relative truth value is not maximal.

(III) Do not take an arbitrary choice between two hypotheses unless (1) their expected epistemic gains are equal and greater than the norm of acceptability, or (2) their expected epistemic gains equal the norm of acceptability and their posterior truth values are maximal.

(IV) A hypothesis of which the expected epistemic gain is or may be smaller than the norm of acceptability is to be rejected.

The norm of acceptability is the liminal value which the expected epistemic gain of the hypothesis must surpass in order to be accepted, except when the expected epistemic gain is maximal. Apparently, the norm of acceptability cannot be negative, otherwise we would have to accept hypotheses implying epistemic losses. Neither is it advisable to choose the value 0 as it is demonstrated by the reasoning situation below.

(i) $P(h_1, k \,\&\, p) < P(h_1, k)$

(ii) $P(h_2, k \,\&\, p) < P(h_2, k)$

(iii) $P(h_1, k \,\&\, p) > P(h_2, k \,\&\, p)$

Let us assume that in one case the reasoners share an orthodox attitude and are not bold. In virtue of the above conditions

$$E_1(h_1, k \,\&\, p) > E_1(h_2, k \,\&\, p)$$
$$E_1(h_1, k \,\&\, p) > 0 \,.$$

According to (I), h_1 should be accepted. But the conditions (i) and (ii) make it clear that the difference between h_1 and h_2 is merely that the attempt to refute the former by tests was less successful.

It implies a similar disadvantage to take any constant number between 0 and 1 as the value norm of acceptability because again the anomaly described by conditions (i)—(iii) would not be totally excluded. Should we choose e.g. the value 1/2, the conditions of acceptability may be fulfilled even though h_1 is partly refuted. At the same time, any value between 0 and 1 may prove to be too strict a norm of acceptability, as exemplified by this situation:

a) $P(h_1, k \,\&\, p) > P(h_1, k) > 0$

b) $1/2 > P(h_1, k \,\&\, p) > P(h_2, k \,\&\, p) \,.$

Let us now suppose that the reasoners possess a heterodox, theoretical, and bold type of approach. They then apply E_3, and

$$E_3(h_1, k \,\&\, p) = P(h_1, k \,\&\, p) - P(h, k) < 1/2 \,.$$

The rejection of h_1 would follow from (IV), though the probability of h_1 may have been considerably increased by tests and it is compatible with the heterodox, theoretical and bold approach to accept every hypothesis which has its probability increased by tests even at a very small rate. It is even less warranted to choose 1 as the norm of acceptability, for on the basis of E_3 it would become impossible to accept any hypotheses.

Therefore it is more appropriate to determine as the norm of acceptability a function that reflects the attitudes of reasoners. The expected epistemic utility of the background knowledge k seems to be most suitable, since k equals for every function E, $1-\lambda$, that is, the degree of caution σ. Our choice is also justified by the fact that a hypothesis can only be rationally accepted if its expected epistemic gain is greater than the degree of caution. Only in this case can the hypothesis be said to have an effect of increasing knowledge. However, if the given function E assumes its highest possible value in the course of justifying the hypothesis tested, this value must be regarded as sufficient for accepting the hypothesis even if it just equals the degree of caution.

In virtue of these considerations and of principles (I)—(IV), we may formulate the following rules of acceptance:

(Ac1)   The hypothesis h_1 is accepted as against h_2, in virtue of (I), if

    a) $E(h_1, k \& p) \geq E(h_2, k \& p)$ and $E(h_1, k \& p) > 1-\lambda$ or

    b) $E(h_1, k \& p) \geq E(h_2, k \& p)$ and $P(h_1, k \& p) = 1$.

(Ac2)   Both the acceptance and the rejection of h_1 and h_2 are suspended according to (II) if

$$E(h_1, k \& p) = E(h_2, k \& p) = 1-\lambda \quad \text{and}$$

$$P(h_1, k \& p) \neq 1, \; P(h_2, k \& p) \neq 1.$$

(Ac3)   Choice between h_1 and h_2 is arbitrary according to (III) if

    a) $E(h_2, k \& p) = E(h_2, k \& p) > 1-\lambda$ or

    b) $P(h_1, k \& p) = P(h_2, k \& p) = 1$.

(Ac4)   Both hypotheses h_1 and h_2 are rejected on the basis of (IV)
             if

        a) $E(h_1, k \,\&\, p) < 1 - \lambda$

        and

        b) $E(h_2, k \,\&\, p) < 1 - \lambda$.

E can be concretized with the help of E_1, E_2, E_3, and E_*, and in the case of E_1, we can assume that $k \,\&\, p \equiv k$.

In the forthcoming, it will be examined how to apply the above rules of acceptance to different situations of reasoning. Such concrete modes of application will be termed *reasoning strategies* because they supply procedures to solve certain tasks of reasoning.

13 REASONING STRATEGIES

Consider a synchronic reasoning situation which is described by $V_S'1$. Assume the hypotheses h_1 and h_2, and that h_2 is a logical consequence of h_1, and also that h_1 is not precluded on the basis of k. Correspondingly, the relative truth of h_2 on the basis of k is at least as great as that of h_1, i.e. $P(h_2, k) \geq P(h_1, k)$. Hence $P(h_2, k) > 0$. In synchronic inferences it is of no significance whether the arguers' attitudes are empirical or theoretical since $k \equiv k \& p$ makes the values of every function E equal $P(h, k)(1 - \lambda)$, i.e. $E_1(h, k)$. Accordingly it must be assumed that $\lambda \neq 1$. In virtue of (Ac1) the condition of acceptance in the given case is that $E(h_2, k) = 1 - \lambda$. This condition would only be fulfilled if $P(h_2, k)$ equalled 1, which is neither precluded nor certain. Since there is no way to abstain, h_2 must be rejected on the basis of (Ac2). It is a different case if h_1 is certain on the basis of k. Then according to V_S1, $P(h_2, k) = 1$, and h_2 must be accepted on the basis of (Ac1). That is, the reasoning situation represented by V_S1 only allows for acceptance while $V_S'1$ must be declared worthless from the point of view of reasoning. The same may be said of other inferences of the V_S and V_S' types.

Consider next the situation of reasoning in which hypotheses are subjected to prior and posterior valuation, i.e. to which a diachronic approach is applied. Let us take as our point of departure the most complicated situation of reasoning, characterized by the function E_*.

The function E_* represents the continuity and discontinuity of cognition at the same time, as well as the unity of the empirical and the theoretical approaches and is compatible with any degree of boldness. It can take account of all types of the probabilistic inferences here discussed, except those of the type V_S'.

By joining in the rules of acceptance (Ac1)—(Ac4) with the function E_* we get the rules of acceptance (Ac$_*$1)—(Ac$_*$4). These rules are considered to be the most general strategies of acceptance.

Examine the case when $\alpha = 1/2$ and $\lambda = 1/2$. Then

$$E_*(h, k \,\&\, p) = P(h, k \,\&\, p) - \frac{1}{4} P(h, k \,\&\, p) + P(h, k).$$

The condition on which to accept the hypothesis h_1 as against h_2 in virtue of (Ac$_*$1) is that

$$E_*(h_1, k \,\&\, p) \geq E_*(h_2, k \,\&\, p)$$

$$E_*(h_1, k \,\&\, p) > 1/2.$$

It can be shown that the conditions are fulfilled if

$$\frac{P(h_1, k \,\&\, p) - P(h_2, k \,\&\, p)}{P(h_1, k) - P(h_2, k)} \geq 1/3$$

and

$$P(h_1, k \,\&\, p) > \frac{2 + P(h_1, k)}{3}.$$

This means that the difference between the posterior relative truth values of h_1 and h_2 is greater than, or equal to, one third of the difference between their prior relative truth values, and that the posterior relative truth value of h_1 is higher than $2/3$.

We can nevertheless introduce stricter requirements. Let us assume that the arguers' attitudes are wholly empiricistic, i.e. $\alpha = 0$. Then

$$E_*(h, k \,\&\, p) = E_2(h, k \,\&\, p),$$

that is, the wholly empiricistic attitude is orthodox at the same time, and it is also readily noticed that it is incompatible with the bold behaviour since in this case $1 - \lambda = 0 = E_2(h, k \,\&\, p)$.

It may appear surprising at first sight that in science, radical empiricism should lead to orthodoxy and vice versa. But if we think that heterodoxy consists in producing new ideas, not new facts, this strategy already appears less one-sided.

A researcher may also be a conformist if his attitude is radically theoretical or empirical-theoretical, provided he is cautious i.e., $\lambda=0$. In this case $E_3(h, k \,\&\, p)=1$, and $E_*(h, k \,\&\, p)=1$. The hypothesis h can only be accepted if $P(h, k \,\&\, p)=1$, that is, if the conforming thinker with his theoretical or empirical-theoretical attitude practically behaves as if his attitude were empiricistic and orthodox.

The most liberal approach is characteristic of heterodox and bold arguers, who always share a theoretical attitude, too. Inferences of the type V'_D can only be applied within the frame of this approach without any restrictions. Then for the hypothesis h_1 to be accepted as against h_2, the following conditions must be satisfied:

$$\frac{P(h_1, k \,\&\, p) - P(h_2, k \,\&\, p)}{P(h_1, k) - P(h_2, k)} \geq 1$$

$$P(h_1, k \,\&\, p) > P(h_1, k).$$

If, however, $0 < \lambda < 1$, the conditions are altered. Let us examine the case when $\lambda = 1/2$. Now the first condition is somewhat moderated, but the second becomes stricter:

$$\frac{P(h_1, k \,\&\, p) - P(h_2, k \,\&\, p)}{P(h_1, k) - P(h_2, k)} \geq 1/2$$

$$P(h_1, k \,\&\, p) > \frac{1 + P(h_1, k)}{2}.$$

On the analogy of (Ac_*1)—(Ac_*4), rules of acceptance corresponding to E_2 and E_3 could also be formulated. This may be done by the reader without difficulty on the basis of the foregoing. We note, however, that the application of E_2 leaves no place for abstention since equality with $1 - \lambda$ as a maximal expected epistemic gain is a condition of acceptance. The same is the case if E_3 is applied with a cautious behaviour.

In the light of the reasoning strategies discussed, let us examine some concrete, though of course somewhat idealized, reasoning situations.

14 THREE EXAMPLES

(1) R. Descartes assigned our innate ideas a rather significant role in cognition. This conception has essentially been revived in Noam Chomsky's theory called Cartesian linguistics, in an attempt to solve the problem of language acquisition (Chomsky, 1967, 2—11). Let us simply refer to this theory as the "hypothesis of innate ideas" and denote it with an h. A more detailed but still sketchy outline of its contents might run like this: small children acquire a language quickly and far more precisely than adults do, often without any external instruction and reinforcement, and almost independently of their intelligence. Let p stand for these empirical statements concerning language acquisition. Chomsky suggests that perhaps one tenth of mankind would only be able to acquire a language were it not for the truth of the innateness hypothesis. Accordingly, his basic assumption is that a small child can acquire a language quickly, precisely, and independently of his intelligence if, and only if, certain structures of natural language are already programmed into man at the moment of his birth. As follows from what has been said,

a) $P(h \equiv p, k) = 1$

b) $P(h, k) > 0$.

In virtue of T17, it follows from a) and b) that $P(h, k) = P(p, k)$ and, on the basis of $V_S 1$, $P(h, k \,\&\, p) = 1$ (considering that, when $P(h \equiv p, k) = 1$, $P(p \supset h, k) = 1$). In this case the hypothesis h can be accepted on any strategy. But the truth of the condition a) remains questionable.

H. Putnam claims that the use of "if, and only if" is unjustified and he allows only an "if. . . , then" relation (Putnam, 1967, 12—22), i.e.:

$$\text{a')} \quad P(h \supset p, k) = 1.$$

It is universally agreed that the "hypothesis of innate ideas" cannot itself be regarded as unscientific, thus

$$\text{b)} \quad P(h, k) > 0,$$

nor is the conclusion p considered as certain on the basis of the background knowledge:

$$\text{c)} \quad P(p, k) < 1.$$

According to $V'_S 1$, $P(h, k) \leq P(p, k)$, and from $V_D 1$, $P(h, k \,\&\, p) > P(h, k)$. If there are no other requirements fulfilled, this conclusion can only be accepted if the arguers employ a theoretical, heterodox, and bold strategy, i.e. no more is required than that

$$P(h, k \,\&\, p) - P(h, k) > 0.$$

Note that here $\sim h$ is considered the rival hypothesis to h. Therefore, for the acceptance of h it is also required that $E_3(h, k \,\&\, p) > E_3(\sim h, k \,\&\, p)$. This condition is equivalent to $P(h, k \,\&\, p) > 0$ when $\lambda = 1$, which is automatically fulfilled on the basis of the conclusion $P(h, k \,\&\, p) > P(h, k) > 0$.

It is also possible to furnish neologistic and moderate strategies which allow for the acceptance of h. Here is one example.

$$\text{a')} \quad P(h \supset p, k) = 1$$

$$\text{b')} \quad P(h, k) > 1/2$$

$$\text{c')} \quad P(p, k) < 6/10$$

$$\text{d)} \quad \alpha = \lambda = 1/2.$$

In the given case, $E_*(h, k \,\&\, p) > 1/2$ and therefore h is acceptable. In the frame of a comparative approach, however, c') can hardly be expected to be fulfilled. Nevertheless, it can be stated that it is hard to imagine a quick, precise acquisition of speech, practically independently of the intelligence, without some genetic program, even though there are other (mainly psychological and social) factors playing an important part too. So we may be justified to suppose $P(p, k)$ not much greater than $P(h, k)$.

(2) A well-known polemic took place at the beginning of the 19th century between advocates of caloric theory (assuming a hypothetic matter of heat) and proponents of the hypothesis of thermal motion. The caloric hypothesis assumed that thermal phenomena were caused by a special material which mixed with bodies: the more caloric a body contained, the hotter it was supposed to be. Caloric had the capacity to flow from hotter bodies into colder ones. Thus the caloric hypothesis was able to explain bodies absorbing heat by convection. However, it could not properly account for heat generated by friction. Supporters of the hypothesis proposed that the thermal capacities of bodies might be reduced by friction, which caused them to be heated even though the quantities of caloric contained in them had not changed. Almost simultaneously with the caloric hypothesis, the hypothesis of thermal agitation emerged in an attempt to explain thermal phenomena by the kinetic motion of the molecules. This view holds that the more intense the agitation of the molecules making up a body, the hotter the body itself. Davy refuted the caloric hypothesis and supplied crucial evidence in favour of thermal agitation by an experiment in which he rubbed two lumps of ice to each other. The ice melted, even though it was impossible for it to take over any caloric from some warmer body. Then he melted another lump of ice of the same mass over fire. The resulting water was in no way different from the one obtained by melting ice by rubbing.

The reasoning situation in question can be characterized by the following premisses:

Certain: the source of thermal phenomena cannot at the same time be a special kind of matter (caloric) and the thermal agitation of molecules.

Certain: if the source of thermal phenomena is some caloric matter, ice can only be melted by contact with a warmer (e.g. burning) body.

Certain: if the source of thermal phenomena is not the motion of molecules, ice can only be melted by contact with a warmer body.

The prior relative truth values of the hypothesis that thermal phenomena come from caloric and of the one that they come from the agitation of molecules are equal and not minimal.

As it has turned out that ice is not only melted by contact with some warmer body but by rubbing too, it is impossible for the source of thermal phenomena to be some special kind of matter (caloric), and they are sure to be caused by the agitation of molecules.

Let h_1 stand for the caloric hypothesis, h_3 for the hypothesis of thermal motion, and h_2 for the melting of ice by heating. Then the reasoning under discussion corresponds to V_D9.

(3) Now let us consider a pattern of reasoning in the field of history. The majority of Hungarian historians share the view that Hungarians, led by their Prince Árpád, settled in the Carpathian Basin about the year 896. Let this view be termed the hypothesis of "single conquest" and denoted by h_1. On the other hand, there are also eminent historians who suppose that the Hungarians first entered that territory towards the end of the 7th century and that the so-called late Avars were also Hungarians. Of course, by this they do not mean to deny that the majority of the Hungarians arrived in their new country as late as 896. We shall call this theory the hypothesis of "double conquest" and denote it by h_2. Let p denote the statement that the Avars disappeared from the Carpathian Basin at the end of the 9th century and that there have remained no Avar place-names to attest to the relatively long stay of that people in the area. (Cf. Gy. László, 1970, 161—190). The hypothesis of the "double conquest" is commonly held to be less probable than that of the "single conquest", moreover, the additional

information p is quite improbable on the basis of k. Assume also that $P(h_2, k) > 0$. On these grounds, the premisses of the argument are as follows:

a) $P(h_1 | h_2, k) = 1$

b) $P(\sim h_1 \supset p, k) = 1$

c) $0 < P(h_2, k) < P(h_1, k) < 1$

d) $P(p, k) \ll 1$.

From a) and b), in virtue of V_S5, it follows that $P(h_2 \supset p, k) = 1$. This conclusion, using c) and d) according to V_D1 yields that $P(h_2, k \& p) \gg$ $\gg P(h_2, k)$. Similarly, b), c), and d) lead to $P(\sim h_1, k \& p) \gg P(\sim h_1, k)$, transcribed as $P(h_1, k \& p) \ll P(h_1, k)$. From c) and d) it is seen that $P(h_1, k) \gg P(h_2, k)$. As a result, the hypothesis of "single conquest", earlier attributed almost complete certainty, should now be rejected and the hypothesis of "double conquest" accepted, inspite of the latter's rather low initial probability. We must note, however, that adherents of the "single conquest" hypothesis generally do not accept the additional knowledge p as proved and persist in their original position.

15 AGREEMENT IN CASE OF COGNITIVE DISSONANCE

Consider a cognitive situation in which scientists F_1 and F_2 must decide on accepting or not the hypothesis h if k is not wholly excluded and it has been possible to justify a conclusion p of h such that is not completely certain on the basis of k. Assume that F_1 has the epistemic attitudes α_1 and λ_1, while F_2 has α_2 and λ_2, and $\alpha_1 = \alpha_2 = 1$ and $\lambda_1 \neq \lambda_2$.

According to the conditions given, and in virtue of $V'_D 1$, $P(h, k \& p) > > P(h, k)$. Since $\alpha_1 = \alpha_2 = 1$, h is attributed the expected epistemic gain $E_3^1(h, k \& p)$ by F_1, and $E_3^2(h, k \& p)$ by F_2, for which it holds that

$$(24) \qquad E_3^1(h, k \& p) = P(h, k \& p) - \lambda_1 P(h, k)$$

$$E_3^2(h, k \& p) = P(h, k \& p) - \lambda_2 P(h, k).$$

Agreement between F_1 and F_2 upon the acceptance of h is achieved if

$$(25) \qquad E_3^1(h, k \& p) > 1 - \lambda_1$$

$$E_3^2(k, k \& p) > 1 - \lambda_2.$$

From (24) and (25),

$$(26) \qquad \frac{P(\sim h, k \& p)}{P(\sim h, k)} < \lambda_1$$

$$\frac{P(\sim h, k \& p)}{P(\sim h, k)} < \lambda_2.$$

It is obvious from the inequalities in (26) that F_1 and F_2 can only

agree upon accepting h if the novelty-value quotient $\dfrac{P(\sim h, k \,\&\, p)}{P(\sim h, k)}$
is smaller than the lower one of the degrees of boldness λ_1 and λ_2, i.e.

$$(27) \qquad \frac{P(\sim h, k \,\&\, p)}{P(\sim h, k)} < \min(\lambda_1, \lambda_2).$$

An interpretation of (27) may run like this: If hypothesis h is acceptable on the basis of its justified consequence p for the more precautious one of two scientists, then there is agreement between them as to its acceptability.

Naturally enough, they cannot both abstain in the given cognitive situation, yet they may agree upon rejecting h, a condition of which is that

$$(28) \qquad \frac{P(\sim h, k \,\&\, p)}{P(\sim h, k)} > \max(\lambda_1, \lambda_2).$$

(28) means that, even though $p's$ consequence has been justified, both F_1 and F_2 agree on rejecting h because it is not even acceptable for the bolder scientist.

Let us assume that $\alpha_1 = \alpha_2 = 0$ and $\lambda_1 \neq \lambda_2$. Then F_1 and F_2 employ the function E_2 since their empirical approach makes them orthodox, that is,

$$E_2^1(h, k \,\&\, p) = (1 - \lambda_1)P(h, k \,\&\, p)$$

$$E_2^2(h, k \,\&\, p) = (1 - \lambda_2)P(h, k \,\&\, p).$$

We have seen that, in the case of E_2, the condition of acceptance is that $E_2(h, k \,\&\, p) = 1 - \lambda_1$ which, provided that $\lambda \neq 1$, is only fulfilled if $P(h, k \,\&\, p) = 1$. If $\lambda = 1$, it will make any h acceptable. This case, however, must be excluded as unreasonable. Therefore, if $\lambda_1 \neq 1$ and $\lambda_2 \neq 1$, then F_1 and F_2 cannot agree upon acceptance unless the hypothesis h is certain on the basis of k and p. In any other case they agree upon rejecting it.

Let us generalize the two cases discussed by only requiring that $\alpha_1 = \alpha_2 = \alpha$. Now the degree of acceptability is to be determined on the basis of the function E_*, according to the equalities

$$(29) \quad E_*^1(h, k \& p) = (1-\alpha)(1-\lambda_1)P(h, k \& p) + \alpha(P(h, k \& p) -$$
$$- \lambda_1 P(h, k)) > 1 - \lambda_1$$
$$E_*^2(h, k \& p) = (1-\alpha)(1-\lambda_2)P(h, k \& p) + \alpha(P(h, k \& p) -$$
$$- \lambda_2 P(h, k)) > 1 - \lambda_2.$$

It follows from the equalities (29) that

$$(30) \qquad \frac{1-\lambda_1}{\alpha\lambda_1} < \frac{P(h, k \& p) - P(h, k)}{1 - P(h, k \& p)}$$

$$\frac{1-\lambda_2}{\alpha\lambda_2} < \frac{P(h, k \& p) - P(h, k)}{1 - P(h, k \& p)}$$

F_1 and F_2 can agree upon accepting h on the basis of k and p if, and only if,

$$(31) \qquad \max\left(\frac{1-\lambda_1}{\alpha\lambda_1}, \frac{1-\lambda_2}{\alpha\lambda_2}\right) < \frac{P(h, k \& p) - P(h, k)}{1 - P(h, k \& p)}$$

It can be shown that, when $\lambda_1 < \lambda_2$, $\dfrac{1-\lambda_1}{\alpha\lambda_1} > \dfrac{1-\lambda_2}{\alpha\lambda_2}$ $\alpha \neq 0$. Correspondingly, F_1 and F_2 can agree upon acceptance if, and only if, h is acceptable for the more precautious F_1 too, on the basis of k and p.

Again there is no way to abstain, but rejection is possible when

$$(32) \qquad \min\left(\frac{1-\lambda_1}{\alpha\lambda}, \frac{1-\lambda_2}{\alpha\lambda}\right) > \frac{P(h, k \& p) - P(h, k)}{1 - P(h, k \& p)}$$

that is, h is not even acceptable for the bolder scientist on the basis of k and p.

Finally, let us examine the case when no requirements are set concerning the attitudes of F_1 and F_2. We get the conditions of mutual acceptance and rejection by generalizing (31) and (32):

$$(33) \qquad \max\left(\frac{1-\lambda_1}{\alpha_1\lambda_1}, \frac{1-\lambda_2}{\alpha_2\lambda_2}\right) < \frac{P(h, k \& p) - P(h, k)}{1 - P(h, k \& p)}$$

96

$$(34) \quad \min\left(\frac{1-\lambda_1}{\alpha_1\lambda_1}, \frac{1-\lambda_2}{\alpha_2\lambda_2}\right) > \frac{P(h, k\,\&\,p) - P(h, k)}{1 - P(h, k\,\&\,p)}.$$

In the case under duscussion, agreement is possible upon abstention, too, achieved when the condition

$$(35) \quad \frac{1-\lambda_1}{\alpha_1\lambda_1} = \frac{1-\lambda_2}{\alpha_2\lambda_2} = \frac{P(h, k\,\&\,p) - P(h, k)}{1 - P(h, k\,\&\,p)}$$

is fulfilled. Here the different epistemic attitudes complement each other, as it were, like when the resultants of two different pairs of forces are equal.

Even more complex situations of acceptance arise when F_1 and F_2 have to choose between two or more hypotheses. The rationality of the decision is then ensured by the combination of (Ac.1)—(Ac.4) and the conditions (33)—(35). Let (Ac'.1)—(Ac'.4) denote the resulting rules of decision.

(Ac'.1)   Scientist F_1 with an attitude α_1 and λ_1 agrees with scientist F_2 with an attitude α_2 and λ_2 upon accepting the hypothesis h_1 as against h_2 if, and only if

a) $E_*^1(h_1, k\,\&\,p) \geq E_*^1(h_2, k\,\&\,p)$ and $E_*^1(h_1, k\,\&\,p) > 1-\lambda_1$

      $E_*^2(h, k\,\&\,p) \geq E_*^2(h_2, k\,\&\,p)$ and $E_*^2(h_1, k\,\&\,p) > 1-\lambda_2$

      and

$$\max\left(\frac{1-\lambda_1}{\alpha_1\lambda_1}, \frac{1-\lambda_2}{\alpha_2\lambda_2}\right) < \frac{P(h_1, k\,\&\,p) - P(h_1, k)}{1 - P(h_1, k\,\&\,p)}$$

      or

b) $E_*^1(h_1, k\,\&\,p) \geq E_*^1(h_2, k\,\&\,p)$ and $P(h_1, k\,\&\,p) = 1$

      and

      $E_*^2(h_1, k\,\&\,p) \geq E_*^2(h_2, k\,\&\,p)$ and $P(h_1, k\,\&\,p) = 1$

(Ac'.2)—(Ac'.4) can be formulated in a similar way.

As these intricate rules of decision may demonstrate, it cannot always be expected, even in rational i.e. purely idealized scientific reasoning, that arguers agree in their value judgements. Their epistemic attitudes may in fact suggest them to make their value judgements different. However, they can come to agreement even if there are significant differences between their epistemic attitudes, since value judgements are ultimately determined by the epistemic attitudes in conjunction, a possibility to counterbalance their mutual one-sidedness. Even though two arguers may not agree upon their judgements of the cognitive situation, they may still agree on accepting, or rejecting a hypothesis, or on abstaining from either decision. Quite different considerations may readily lead to similar value judgements.

16 SOME OTHER CONCEPTIONS

It is a widespread method of valuation in ball games that, score points being the same, the difference in placing two teams is made by the proportion of scores, or the difference of scores. On this analogy, the acceptability of hypotheses can also be measured by the quotients and by the differences of their posterior and prior probabilities. The former is termed *confirmation quotient* (Rosenboom, 1971, 342—347), the latter *confirmational effect* (von Wright, 1970, 599—600), denoted by the symbols C_q and C_e respectively. The definitions of the confirmation quotient and the confirmational effect are

$$C_q(h, k \,\&\, p) = \frac{P(h, k \,\&\, p)}{P(h, k)} \qquad /P(h, k) \neq 0/$$

$$C_e(h, k \,\&\, p) = P(h, k \,\&\, p) - P(h, k)$$

The functions C_q and C_e assume values within the intervals $[0, \infty]$ and $[1, 1]$ respectively. It can be shown that the confirmational effects are different when the confirmation quotients are equal, i.e.

$$\text{if} \qquad C_q(h_1, k \,\&\, p) = C_q(h_2, k \,\&\, p),$$

$$\text{then} \qquad C_e(h_1, k \,\&\, p) \neq C_e(h_2, k \,\&\, p),$$

provided that $P(h_1, k) \neq P(h_2, k)$. The reverse is also true if the conditions stated are fulfilled.

The functions C_q and C_e also imply the possibility to define rules of acceptance on the scheme of the rules (Ac1)—(Ac4). The liminal value of acceptability is 1 for the former, and 0 for the latter function. These values being constant, they presuppose minimal confirmation and may therefore lead to liberal strategies of acceptance. Note however, that the joint application of the functions C_q and C_e makes it possible to avoid arbitrary choice in cases of equality because e.g., when $C_q(h_1, k \& p) = C_q(h_2, k \& p)$, $C_e(h_1, k \& p) \neq C_e(h_2, k \& p)$ and vice versa.

It is peculiar of the function C_q that it equals $1/P(p, k)$ when $P(h \supset p, k) = 1$, so it is independent of the value of $P(h, k)$ /provided that $P(h, k) \neq 0$/. On the other hand, the value of the function C_e depends on $P(h, k)$ too, therefore, it is a more suitable measure than the function C_q.

It may be recognized that the function C_e is actually the special case of E_3 when $\lambda = 1$, that is, it presupposes a theoretical, heterodox, and bold strategy.

We can also apply the *systematic power* (Pietarinen, 1970, 123—147) to judge acceptability, defined as follows:

$$S_t(h, k \& p) = \frac{P(p, k \& p) - P(p, k)}{1 - P(p, k)}.$$

The systematic power can be recognized as the quotient of the confirmational effects of h upon the statement p expressing empirical knowledge on the one hand, and of p on itself on the other, i.e.,

$$S_t(h, k \& p) = \frac{C_e(p, k \& h)}{C_e(p, k \& p)}.$$

It can be shown that, when $P(p, k \& h_1) > P(p, k \& h_2)$ $S_t(p, k \& h_1) > S_t(p, k \& h_2)$, i.e. the posterior relative truth values of p clearly determine the acceptability of h_1 or h_2 provided that $P(p, k) \neq 0$. But the function S_t only provides grounds for rather weak rules of acceptance since the norm of acceptability must be chosen as a constant value. Because the function S_t varies within the interval $[\sim \infty, 1]$, it seems most suitable to choose as the norm of acceptance the systematic power of p exerted upon itself, which equals 0.

There is not a very significant difference between the function S_t and the *explanatory power* proposed by Popper (Popper, 1959, 400—402):

$$Expl\,(h, k \& p) = \frac{P(p, k \& h) - P(p, k)}{P(p, k \& h) + P(p, k)}.$$

A more suitable means of determining strategies of acceptance is the degree of corroboration or verisimilitude, also put forward by Popper (Popper, 1959, *loc. cit.*):

$$Corr\ (h, k\ \&\ p) = \frac{P(p, k\ \&\ h) - P(p, k)}{P(p, k\ \&\ h) - P(h\ \&\ p, k) + P(p, k)}.$$

The hypothesis h_1 can be justified to have a higher degree of corroboration than that of h_2 if it grounds the statement p better and its prior probability is at least as high as that of h_2. If they corroborate the statement p to the same degree, then the hypothesis possessing the lower prior probability will have the higher degree of corroboration.

Thus the function *Corr* gives rise to more adequate strategies of acceptance than do the functions S_t or *Expl*. There is still a need to clarify the difference between C_e and *Corr*.

Assume that the hypotheses h_1 and h_2 are being tested on the basis of the empirical statement p, and that $P(h_1, k) = P(h_2, k) \neq 0$ and $P(h_1, k\ \&\ p) > P(h_2, k\ \&\ p) > 0$.

Then $\quad P(h_1, k\ \&\ p) - P(h_1, k) > P(h_2, k\ \&\ p) - P(h_2, k), \quad$ i.e. $C_e(h_1, k\ \&\ p) > C_e(h_2, k\ \&\ p)$.

Let us compute the corresponding *Corr* values, too. To make that operation simpler, let us transform the inequality

$$\frac{P(p, k\ \&\ h_1) - P(p, k)}{P(p, k\ \&\ h_1) - P(h_1\ \&\ p, k) + P(p, k)} >$$
$$> \frac{P(p, k\ \&\ h_2) - P(p, k)}{P(p, k\ \&\ h_2) - P(h_2\ \&\ p, k) + P(p, k)}$$

into

$$P(p, k\ \&\ h_1)\,[2 - P(h_1 k)] > P(p, k\ \&\ h_2)\,[2\ \ P(h_2, k)]$$

since

$$P(p, k\ \&\ h) = \frac{P(p, k) \cdot P(h, k\ \&\ p)}{P(h, k)}$$

after the necessary substitutions and reductions, we get

$$P(p, k\ \&\ h_1) > P(p, k\ \&\ h_2).$$

Making use of this inequality and the premiss $P(h_1, k) = P(h_2, k)$, we obtain the result $Corr(h_1, k \& p) > Corr(h_2, k \& p)$. The result is the same if we suppose that $P(h_1, k \& p) = P(h_2, k \& p)$ and $P(h_1, k) < < P(h_2, k)$. In general, however, it does not obtain that, if $C_e(h_1, k \& p) > > C_e(h_2, k \& p)$ then $Corr(h_1, k \& p) > Corr(h_2, k \& p)$. Thus e.g., if $P(h_1, k \& p) = 2P(h_1, k)$, $P(h_2, k \& p) = 2P(h_2, k)$ and $P(h_1, k) < P(h_2, k)$, then $Corr(h_1, k \& p) < Corr(h_2, k \& p)$.

The question arises whether to consider the confirmational effect or the degree of corroboration as the more adequate measure of acceptability.

The degree of corroboration seems to be more adequate because it not only takes account of the relation between the differences $P(h_1, k \& p) - P(h_1, k)$ and $P(h_2, k \& p) - P(h_2, k)$ but also of the order of magnitude of $P(h_1, k \& p)$ and $P(h_2, k \& p)$. This is indicated by the fact that the inequality

$$Corr(h_1, k \& p) > Corr(h_2, k \& p)$$

can be transformed into

$$\frac{P(h_1, k \& p) - P(h_1, k)}{P(h_1, k \& p) - P(h_1, k)P(\sim h_1, k \& p)} >$$
$$> \frac{P(h_2, k \& p) - P(h_2, k)}{P(h_2, k \& p) - P(h_2, k)P(\sim h_2, k \& p)}$$

by a brief calculation. Hence

$$\frac{C_e(h_1, k \& p)}{P(h_1, k \& p) - P(h_1, k)P(\sim h_1, k \& p)} >$$
$$> \frac{C_e(h_2, k \& p)}{P(h_2, k \& p) - P(h_2, k)P(\sim h_2, k \& p)}.$$

That is, according to *Corr*, we are to accept the hypothesis on which the statement p exerts a greater effect of confirmation than on its rival, and which is at the same time more probable when tested than its rival.

Nonetheless it must be noted that a fixed norm of acceptance is also presupposed by the function Corr, which again makes its adequacy rather questionable.

Finally there is another measure, proposed by N. Rescher: the *degree of grounding*, which may be utilized in evaluating the acceptability of hypotheses (Rescher, 1970, 72—88). Let us denote it by S_p and define it as follows:

$$S_p(h, k \mathbin{\&} p) = \frac{P(h, k \mathbin{\&} p) - P(h, k)}{1 - P(h, k)} \cdot P(p, k).$$

S_p varies within the closed interval $[-1, 1]$, its value being 0 if the value of $P(h, k \mathbin{\&} p) - P(h, k)$ and/or $P(p, k)$ is 0. It is required that $P(h, k) \neq 1$.

As it is easily recognized, the direction of the variation of S_p is identical with that of C_e, therefore it provides no grounds for the formulation of new rules of acceptance.

As a result of this brief review, we may point out that the function suitable to formulate the most adequate rules of acceptance is the function E_*, since it alone can fully account for all the important factors in the field of reasoning. E_1, E_2, and E_3 are special cases of E_*. The rules of acceptance according to C_e are in fact the rules corresponding to E_3, provided that $\lambda = 1$. The other functions examined are no more adequate for the acceptance of hypotheses than C_e is, except for the function Corr, though the latter is still not as adequate as E_*. Therefore it is most reasonable to use the functions E in evaluating the acceptability of hypotheses. The nature of the reasoning situation will decide which of these functions is the most suitable for the given case.

17 VERITAS FILIA TEMPORIS

We have seen that rational (i.e. rather idealized) scientific reasoning depends on quite a few factors, and presupposes the homogeneity of the field of reasoning. However, scientific life does not usually fulfil this condition perfectly. It can scarcely be expected that scientists in different historical ages are, like completely rational beings, strictly guided by the distant goals of cognition and by values adequately reflecting the reasoning situations, when they take their decisions following methodological rules resulting from the generalization of the historical progress of scientific knowledge. If the background knowledge of the arguers is the same, if they agree upon their judgements of the truth values of the hypotheses discussed, and if they apply the methods of inference justified by probabilistic logic, then the field of reasoning is homogeneous enough to ensure the same decisions even in case of cognitive dissonance. Naturally enough, this requires different epistemic attitudes to complement each other. Then it may easily happen that, even though in agreement, both arguers are wrong. Mistaken value judgements are none the less somehow levelled out in the long run, and time is on the side of truth. An important part is played by the succession of scientific generations, a favourable political and cultural athmosphere, and the widest possible international communication of science. Although such factors seem external from the point of view of science, they exercise a positive influence on its development and thus on the achievement of its cognitive aims.

APPENDIX

1 THE FORMULATION OF PLAUSIBLE INFERENCES IN ALETHIC MODAL LOGIC

It has been pointed out that such terms encountered in plausible inferences as "necessary" ("certain"), "possible", "impossible" ("precluded"), "contingent", etc. are of a *modal* character. As they refer to hypotheses (statements), they can be regarded as relative truth values or plausibilities. The modal logics that explicate modes of truth are called *alethic modal logics*. A precise exposition of these modal logics (Hughes—Cresswell, 1968) cannot be the task of the present study, which is focused on certain problems of application only. We merely undertake to expound the most essential parts of the theoretical foundations.

Apart from "it is possible that p" and "it is possible that $\sim p$", we also introduce the modalities "it is necessary that p", "it is impossible that p", "it is contingent that p", and "it is eventual that p", while from classical two-valued logic the categories "true" and "false" are kept and here made to correspond to the phrases "it is probable that p" and "it is probable that $\sim p$" respectively. The relations between these concepts are demonstrated by Figure 10, which essentially corresponds to Figure 1.

Let N stand for "necessary", M for "possible", C for "contingent", and E for "eventual". The diagram helps the formulation of the following relations:

(1)   It is not necessary that $p \Leftrightarrow$ it is possible that non-p, i.e., by symbols:

$$\sim Np \Leftrightarrow M \sim p.$$

(2)   It is necessary that $p \Leftrightarrow$ it is impossible that non-p, i.e.,

$$Np \Leftrightarrow \sim M \sim p.$$

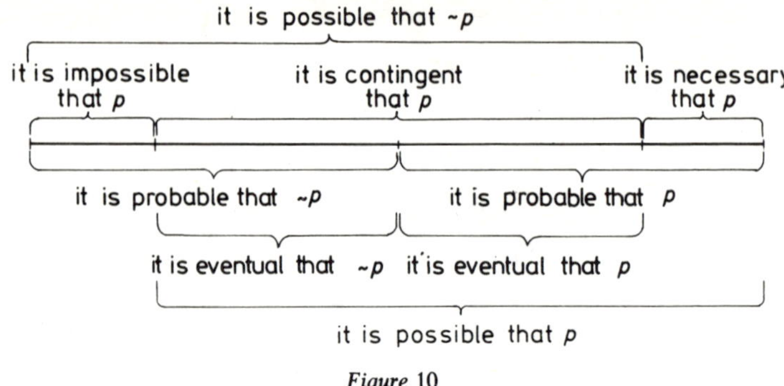

Figure 10

(3)   It is possible that $p \Leftrightarrow$ it is not necessary that non-p, i.e.

$$Mp \Leftrightarrow \sim N \sim p.$$

(4)   It is impossible that $p \Leftrightarrow$ it is necessary that non-p, i.e.

$$\sim Mp \Leftrightarrow N \sim p.$$

(5)   If necessary that p, then p, i.e.

$$Np \Rightarrow p.$$

(6)   If p, then it is possible that p, i.e.

$$p \Rightarrow Mp.$$

(7)   It is contingent that $p \Leftrightarrow$ it is possible that p and it is possible that non-p, i.e.,

$$Cp \Leftrightarrow Mp \;\&\; M \sim p.$$

(8)   It is contingent that $p \Leftrightarrow$ it is contingent that non-p, i.e.

$$Cp \Leftrightarrow C \sim p.$$

(9)   It is eventual that $p \Leftrightarrow$ it is a fact that p, and it is not necessary that p, i.e.:

$$Ep \Leftrightarrow p \;\&\; \sim Np.$$

The following postulates are usually given as well in the construction of alethic modal logic:

(10) $M(p \vee q) \Leftrightarrow Mp \vee Mq$, i.e., that "it is possible that p and/or q" has the same logical value as "it is possible that p and/or it is possible that q". (The principle of M-distribution.)

(11) $M(p \& q) \Rightarrow Mp \& Mq$; if it is possible that p and q, then it is possible that p and it is possible that q.

(12) $N(p \& q) \Leftrightarrow Np \& Nq$; that it is necessary that p and q is logically equivalent with that it is necessary that p and it is necessary that q.

(13) $N(p \supset q) \Rightarrow Np \supset Nq$; the necessity of a conditional implies that, if its major premiss is necessary, so is its minor premiss.

The combination of the laws (1). . . (13) with classical logic yields the simplest theory of normal modal logic. (It is implied by the theory that, if A is a logical truth, NA is one, too.) Note that the list of postulates is redundant; it would suffice to have (3), (5), (7), and (13).

Without writing out the proofs, here are some modal schemes of inference:

(m1) $\{N(p \supset q), Np\} \Rightarrow Nq$

(m2) $\{N(p \supset q), N \sim q\} \Rightarrow N \sim q$

(m3) $\{N(p \vee q), N \sim p\} \Rightarrow Nq$

(m4) $\{N(p | q), Np\} \Rightarrow N \sim q$

(m5) $\{N(p \supset q), N(q \supset r)\} \Rightarrow N(p \supset r)$

(m1)—(m5) may be made to correspond to the schemes of inference $S1$—$S5$.

These modal inferences are also valid:

(m'1) $\{N(p \supset q), Mp\} \Rightarrow Mq$ (m''1) $\{M(p \supset q), Np\} \Rightarrow Mq$

(m'2) $\{N(p \supset q), M \sim p\} \Rightarrow M \sim p$ (m''2) $\{M(p \supset q), \sim Mq\} \Rightarrow M \sim p$

$(m'3)$ $\{N(p \lor q), M \sim q\} \Rightarrow Mp$ $\qquad$ $(m''3)$ $\{M(p \lor q), \sim Mp\} \Rightarrow Mq$

$(m'4)$ $\{N(p|q), Mp\} \Rightarrow M \sim q$ $\qquad$ $(m''4)$ $\{M(p|q), Np\} \Rightarrow M \sim q$

$(m'1)$—$(m'4)$ correspond to $S'1$—$S'4$. Certain synchronic plausible inferences can be brought into correspondence with $(m''1)$—$(m''4)$ as well. These have not been discussed until now.

$S''1$	$p \supset q$	is possible	$S''3$	$p \lor q$	is possible
	p	is certain		p	is precluded
	q	is possible		q	is possible
$S''2$	$p \supset q$	is possible	$S''4$	$p|q$	is possible
	q	is precluded		p	is certain
	$\sim p$	is possible		$\sim p$	is possible.

2 POSSIBILITIES OFFERED BY MANY-VALUED LOGIC

In the following, we shall seek to establish what possibilities are offered by many-valued logic to deal with plausible inferences. Obviously, we cannot undertake to examine each of the great number of existing many-valued logical systems. We confine ourselves to some representative examples to demonstrate that many-valued logical systems may prove suitable frames of plausible inference.

a) Łukasiewicz' generalized many-valued logic ($Ł_n$)

$Ł_n$ (cf. Rescher, 1969, 36—43) is interpreted as follows.

Let $n \geq 2$ be an integer. Divide the numerical interval delimited by 0 and 1 into $n-1$ (equal) parts. Thus we get n different values including the limit terms. Consider these numbers to be (partial) truth values, 0 representing falsity for certain, 1 standing for truth for certain. Choose a value s_0 different from 0 on the scale. Let the values s conforming to the requirement $s \geq s_0$ be called *designated* values. (When choosing that $s_0 = 1$, the only designated value is 1.) In the figure below, $n = 6$, $s_0 \geq \dfrac{2}{5}$.

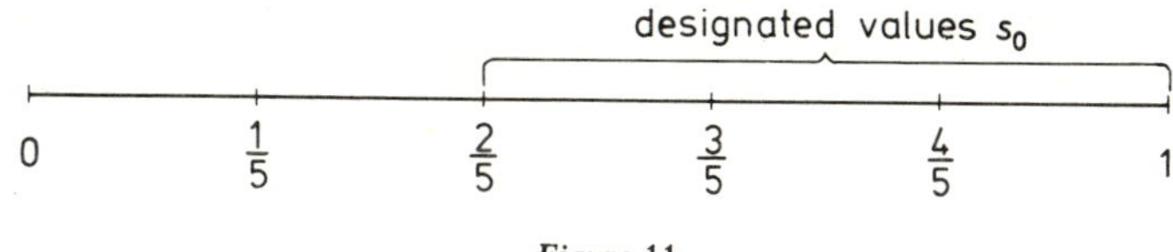

Figure 11

The value s_0 is interpreted as the lower limit of "possible". It is assumed that the statements can possess "relative truth values" according to the scale. The statements of designated values are the more or less acceptable ones. (In the case $n = 2$, we get classical logic.)

Let "$|p|$" denote the value of the statement p. The rules below define ways to calculate the values of compound statements.

(1) $\qquad |{\sim}p| = 1 - |p|$

(2) $\qquad |p \vee q| = \max [|p|, |q|]$

(3) $\qquad |p \,\&\, q| = \min [|p|, |q|]$

(4) $\qquad |p \supset q| = \min [1, 1 - |p| + |q|]$

(5) $\qquad |p \equiv q| = 1 - \||p| - |q|\|$ ('$|\ |$' is the sign for absolute value.)

The following relations are also to be mentioned:

(6) $\qquad |p \vee q| = |(p \supset q) \supset q|$

(7) $\qquad |p \supset q| = |{\sim}({\sim}p \vee {\sim}q)|$

(8) $\qquad |p \supset q| = 1$ if, and only if, $|p| \leq |q|$.

According to the given condition, $|p \supset q| = 1$ since from (4), $|p \supset q| = \min [1, 1 - |p| + |q|]$, and $|1| - |p| + |q| \leq 1$. Thus $|p \supset q| = 1$. Nevertheless, the formula $p \supset q$ only assumes the value 1 if the given condition obtains because, when $|p| > |q|$, $|p \supset q|$ is always smaller than 1.

Now we are presenting some "quasi-rules of inference" valid in n-valued logic. These schemes guarantee the values of conclusions given if the premisses assume the appropriate values.

$$
\begin{array}{lll}
R_L1 & |p \supset q|=1 & \qquad p \supset q \quad \text{is certain} \\
      & |p|=1 & \qquad p \qquad\ \ \text{is certain} \\
\hline
      & |q|=1 & \qquad q \qquad\ \ \text{is certain.}
\end{array}
$$

Thus R_L1 represents $S1$.

$$
\begin{array}{lll}
R_L'1 & |p \supset q|=1 & \qquad p \supset q \quad \text{is certain.} \\
      & |p| \geqq s_0 & \qquad p \qquad\ \ \text{is at least possible} \\
\hline
      & |q| \geqq s_0 & \qquad q \qquad\ \ \text{is at least possible}
\end{array}
$$

As, in case of $|p \supset q|=1$ and in virtue of (8), $|p| \leqq |q|$ and $|p| \geqq s_0$, therefore q is at least as high as s_0.

This corresponds to the plausible scheme $S'1$.

$$
\begin{array}{lll}
R_L2 & |p \supset q| \geqq s_0 & \qquad p \supset q \quad \text{is at least possible} \\
      & |p|=1 & \qquad p \qquad\ \ \text{is certain} \\
\hline
      & |q| \geqq s_0 & \qquad q \qquad\ \ \text{is at least possible.}
\end{array}
$$

If $|p|=1$, then $|p \supset q|=\min[1,|q|]=|q|$. Thus $|q| \geq s_0$, since $|p \supset q| \geq s_0$ according to the assumption.

This rule corresponds to the plausible scheme $S''1$.

A similar treatment can be given in $Ł_n$ of the other synchronic schemes of inference, too.

b) Directional Logic (DL)

Directional Logic has been created by the Polish logician L. S. Rogowski (Rogowski, 1967, 274—307). His aim was to reconstruct Hegel's logical conception by the means of symbolic logic. Directional

Logic is a special kind of four-valued logic, with truth values as follows:

true (1)
nearly true (2/3)
nearly false (1/3)
false (0),

with 2/3 and 1/3 being directed transitional values; 2/3 can be interpreted as approaching truth, and 1/3 as approaching falsehood.

The structure of DL is rather complex. However, it is not necessary here to present the whole of this logic in order to show how certain kinds of plausible inferences can be discussed in it. Therefore, we shall only deal with its features that are necessary for our purpose.

The simplest operation in DL is ordinary negation '$\sim$', defined by the following matrix of values:

p	$\sim p$
1	0
2/3	1/3
1/3	2/3
0	1

In short: $|\sim p| = 1 - |p|$.

There are also in Directional Logic the following negation-like operations: "it starts being the case that" $\vec{N}$ (and "it starts being no longer the case that" $\tilde{N}$). Their value charts:

p	$\vec{N}p$		p	$\tilde{N}p$
1	1/3		1	2/3
2/3	1		2/3	0
1/3	0		1/3	1
0	2/3		0	1/3

In the directional logical formulation of plausible inferences we may use the conditional ($\supset$), conjunction (&), alternation ($\vee$) and incompatibility ($|$). Their value matrices are as follows:

$p \supset q$	1	2/3	1/3	0
1	1	2/3	1/3	0
2/3	1	2/3	1/3	1/3
1/3	1	2/3	2/3	2/3
0	1	1	1	1

$p \,\&\, q$	1	2/3	1/3	0
1	1	2/3	1/3	0
2/3	2/3	2/3	1/3	0
1/3	1/3	1/3	1/3	0
0	0	0	0	0

$p \vee q$	1	2/3	1/3	0
1	1	1	1	1
2/3	1	2/3	2/3	2/3
1/3	1	2/3	1/3	1/3
0	1	2/3	1/3	0

$p \,|\, q$	1	2/3	1/3	0
1	0	1/3	2/3	1
2/3	1/3	1/3	2/3	1
1/3	2/3	2/3	2/3	1
0	1	1	1	1

With the help of the value matrices the following equalities are easily proved:

$$|p \,\&\, q| = \min\left[\,|p|, |q|\,\right]$$

$$|p \vee q| = \max\left[\,|p|, |q|\,\right]$$

$$|p \supset q| = |\sim p \vee q| = \max\left[1 - |p|, |q|\right]$$

$$|p \,|\, q| = |\sim(p \,\&\, q)| = 1 - \min\left[\,|p|, |q|\,\right].$$

Let us correlate the degrees of reliability with certain one-variable operations of directional logic like this:

certain — it is the case that (ordinary assertion)
credible — it starts being the case that
hardly credible — it starts being no longer the case that
precluded — it is not the case that (ordinary negation).

Let us take the following plausible inference:

$$p \supset q \text{ is certain}$$
$$\underline{p \text{ is possible}}$$
$$q \text{ is possible}$$

The correlate in DL of this inference would be

$$p \supset q$$
$$\vec{N}p$$
$$\overline{\vec{N}q} \quad .$$

However, this is not valid according to the value matrices, for with th premises having the value 1 the conclusion takes the value 1/3. Sinc $|\vec{N}q| = 1/3$, $|q| = 1$. This conclusion is, however, too optimistic. Th analogues of S″1 can be formulated:

$$\vec{N}(p \supset q) \qquad \qquad \vec{N}(p \supset q)$$
$$p \qquad \qquad \qquad \vec{N}p$$
$$\overline{\vec{N}q} \qquad \qquad \quad \overline{\vec{N}q} \quad .$$

Let us see the following formula:

$$p \supset q$$
$$\vec{N} \sim q$$
$$\overline{\vec{N} \sim p} \quad .$$

Since $|\vec{N} \sim q| = |\tilde{N}q|$ and $|\vec{N} \sim p| = |\tilde{N}p|$,

$$p \supset q$$
$$\tilde{N}q$$
$$\overline{\tilde{N}p} \quad .$$

Similarly, the formulas are valid:

$$\vec{N}(p \supset q) \qquad \qquad \vec{N}(p \supset q)$$
$$\sim q \qquad \qquad \qquad \tilde{N}q$$
$$\overline{\tilde{N}p} \qquad \qquad \quad \overline{\tilde{N}p} \quad .$$

The following versions of weak reduction are also right:

$$\frac{\begin{array}{c} p \supset q \\ \vec{N}q \end{array}}{\vec{N}p} \qquad\qquad \frac{\begin{array}{c} \vec{N}(p \supset q) \\ \vec{N}q \end{array}}{\vec{N}p}$$

$$\frac{\begin{array}{c} p \supset q \\ \overset{\leftarrow}{N}p \end{array}}{\overset{\leftarrow}{N}q} \qquad\qquad \frac{\begin{array}{c} \vec{N}(p \supset q) \\ \overset{\leftarrow}{N}p \end{array}}{\overset{\leftarrow}{N}} \ .$$

However, the scheme

$$\frac{\begin{array}{c} \vec{N}(p \supset q) \\ q \end{array}}{\vec{N}p}$$

is not valid, for when $|\vec{N}(p \supset q)| = 1$, $|q| \neq 1$.

3 POSSIBILITIES OFFERED BY THE $M_d 1$ CALCULUS

G. H. von Wright has elaborated a so-called *dyadic modal logic*, which displays certain similarities to probabilistic logic (von Wright, 1957, 89—115).

$M_d 1$ is built up of so-called first-order homogeneous propositions which are defined recursively:

(i) The propositional variables p, q, r, and the tautology t, as well as their truth functions, are *0-order M-formulas*.

(ii) The atomic first-order M-formula is of the form $M(\alpha|\beta)$, where α and β are *0-order M-formulas*.

(iii) First-order homogeneous M-formulas are first-order atomic M-formulas, or truth functions of first-order atomic M-formulas.

114

Axioms:

$A1 \qquad M(k|k) \supset \sim M(\sim k|k)$.

Interpretation:
If k is possible on the basis of k (i.e. k is not self-contradictory), then $\sim k$ is impossible on the basis of k.

$A2 \qquad M(p|k) \lor M(\sim p|k)$.

Interpretation:
p is possible and/or $\sim p$ is possible on the basis of k.

$A3 \qquad M(p \,\&\, q|k) \equiv M(p|k) \,\&\, M(q|k \,\&\, p)$.

Interpretation:
That p and q are possible on the basis of k is equivalent with the statement that p is possible on the basis of k and q is possible on the basis of $k \,\&\, p$.

The theorems of the calculus are formulas satisfying the following conditions:

(I) A theorem is derived from tautologies of propositional logic by substituting homogeneous first-order M-formulas for propositional variables.

(II) A theorem is derived from the axioms and/or theorems of the calculus by substituting propositional variables with O-order M-formulas.

(III) A theorem may be the minor term of a conditional which is itself an axiom or a theorem and whose major term is also an axiom or a theorem.

(IV) A theorem may be derived from axioms or theorems of the calculus by substituting O-order M-formulas with tautologically equivalent M-formulas.

Let us introduce the definition

Df. 1 $\qquad N(p|k) = \sim M(\sim p|k)$.

That is, "p is necessary on the basis of k" means the same as "$\sim p$ is impossible on the basis of k".

Employing Df. 1 A1 and A2 can be transcribed as

A1 $\qquad M(k|k) \supset N(k|k),$

that is, if k is compatible with itself, then it is necessary in itself;

A2 $\qquad \sim M(\sim p|k) \supset M(p|k)$ or $N(p|k) \supset M(p|k)$

that is, necessity involves its possibility.

Without their proofs, we present some theorems of $M_d 1$ significant for our purposes:

T1 $\qquad M(p|k) \equiv M(p \& q|k) \vee M(p \& \sim q|k)$

T2 $\qquad M(p \vee q|k) \equiv M(p|k) \vee M(q|k)$

T3 $\qquad N(p \& q|k) \equiv N(p|k) \& N(q|k)$

The following theorems do not appear in von Wright, but they are easily proved:

T4 $\qquad [N(p \supset q|k) \& N(p|k)] \supset N(q|k)$

T5 $\qquad [N(p \vee q|k) \& N(\sim p|k)] \supset N(q|k)$

T6 $\qquad [N(p|q|k) \& N(p|k)] \supset N(\sim q|k)$

T7 $\qquad [N(p \supset q|k) \& N(q \supset r|k)] \supset N(p \supset r|k)$

T8 $\qquad N(p \supset q|k) \equiv [M(p|k) \supset N(q|k \& p)]$

T9 $\qquad N(p \vee q|k) \equiv N(p|k) \vee N(q|k \& \sim p)$

T10 $\qquad N(p|q|k) \equiv M(p|k) \supset N(\sim q|k \& p)$

T11 $\qquad [N(p \supset q|k) \& M(p|k)] \supset N(q|k \& p)$

T12 $\qquad [N(p \supset q|k) \& \sim N(q|k)] \supset N(\sim p|k \& \sim q)$

T13 $\qquad [N(p \vee q|k) \& \sim N(p|k)] \supset N(q|k \& \sim p)$

T14 $\qquad [N(p|q|k) \& M(p|k)] \supset N(\sim q|k \& p)$

T15 $\qquad \{N(p \supset q|k) \& N(q \supset r|k) \& M(p|k)\} \Rightarrow N(r|k \& p).$

116

With the help of T4, the analogues of $S1$ and $S2$ can be grounded in M_d1:

(m_s1) $\qquad \{N(p \supset q \,|\, k), N(p \,|\, k)\} \Rightarrow N(q \,|\, k)$

(m_s2) $\qquad \{N(p \supset q \,|\, k), N(\sim q \,|\, k)\} \Rightarrow N(\sim p \,|\, k)$

T5—T7 make it possible to represent $S3$—$S5$:

(m_s3) $\qquad \{N(p \vee q \,|\, k), N(\sim p \,|\, k)\} \Rightarrow N(q \,|\, k)$

(m_s4) $\qquad \{N(p \,|\, q \,|\, k), N(p \,|\, k)\} \Rightarrow N(\sim q \,|\, k)$

(m_s5) $\qquad \{N(p \supset q \,|\, k), N(q \supset r \,|\, k)\} \Rightarrow N(p \supset r \,|\, k)\,.$

Inferences of the type S'' can also be discussed in M_d1. Their representations are derived from inferences of the type (m_s) by substituting the functor N with M in the second premiss. The conclusions will be weaker accordingly.

The M_d1 calculus is really interesting because it provides a way to represent the diachronic inferences $D1$—$D5$ as well:

On the basis of T11—T15, we get the following schemes:

(m_d1) $\qquad \{N(p \supset q \,|\, k), M(p \,|\, k)\} \Rightarrow N(q \,|\, k \,\&\, p)$

(m_d2) $\qquad \{N(p \supset q \,|\, k), M(\sim q \,|\, k)\} \Rightarrow N(\sim p \,|\, k \,\&\, \sim q)$

(m_d3) $\qquad \{N(p \vee q \,|\, k), M(\sim p \,|\, k)\} \Rightarrow N(q \,|\, k \,\&\, \sim P)$

(m_d4) $\qquad \{N(p \,|\, q \,|\, k), M(p \,|\, k)\} \Rightarrow N(\sim q \,|\, k \,\&\, p)$

(m_d5) $\qquad \{N(p \supset q \,|\, k), N(q \supset r \,|\, k), M(p \,|\, k)\} \Rightarrow N(r \,|\, k \,\&\, p)\,.$

However, the discussion of inferences of the type D' is not even possible in M_d1.

4 THE COMPARISON OF SCIENTIFIC HYPOTHESES

1 EXPOSITION OF THE PROBLEM

Scholars of the theory of science have seriously debated about the development of science in the last two decades. It cannot be the goal of this short appendix to analyze and reconstruct the far-reaching polemies the task of which has been fulfilled by others before us. However, we can not pass by some illuminating lessons of the discussions about the development of science:

a) These polemies have greatly contributed to the disentangling of the problems concerning the development of science and have added viewpoints to the formulation of the solution-alternatives (hypotheses).

b) Generally, these hypotheses have been formulated by purely conceptual devices and so they have never been put to experiential tests.

c) The long delay of the exclusively theoretical debates about the development of science has led many scholars to adopt either a sceptical or a relativistic attitude.

I think that with only the help of the categorial arsenal of the theory of science and that of symbolic logic it is impossible to outline a substantially adequate and formally correct conception of the development of science. We have to make use of such mathematical structures, too, which suitably complement the conceptual and logical analysis.

In the present paper we suggest a positive solution-alternative to the following problem:

(Pr)   How can certain hypotheses suggested for acceptance be compared as answers to well-formed scientific problems?

Before starting to solve the problem (Pr) it seems reasonable, following Descartes' advice, to divide the problem under discussion into the following sub-problems:

(Pr 1)   What sorts of scientific knowledges are comparable?
(Pr 2)   What are those components (factors) on the basis of which the comparison can be made?

(Pr 3)   What function relation holds among the important components?

(Pr 4)   What are the rules of choice that control our decisions concerning the hypotheses suggested for acceptance?

2 COMPARABILITY OF THE HYPOTHESES

The opinion is widely accepted that there are developed and less developed branches of science. Generally (natural) sciences are regarded as more advanced than the social sciences and even within the fields mentioned a certain hierarchy can be set up; physics has advantage over biology, economics over history, etc.

Representatives of those branches of science for whom the comparison proves to be disadvantageous question the very legitimacy of the comparison. They think that any comparison is one-sidedly methodology-centric and the complexity and the problems of metrization of the fields of reality to be examined are completely passed by.

Metaphorically speaking, we cannot demand the same sports results from those in cross-country running as from sportsmen competing under the ideal conditions of a stadium.

Supporters of the unlimited comparison argue in the following way: for about the last hundred years mathematization has penetrated into such fields where measurements and exact experiments have been long held to be too venturesome. This process can be demonstrated by the spread of exact methods in biology, psychology, economics, linguistics, etc. Sciences researching into the material processes have obviously more favourable research conditions than the disciplines exploring the mental life of society but this is no less an objectively given characteristic than, referring back to our metaphor, the physical qualities of the sports ground. Consequently, there will always be more developed and less developed sciences.

We are not going into details concerning the relevance of the different arguments; we say only that neither the concessive nor the refusing attitude towards the comparison seem convincing. If we accepted the presupposition that only the results of the same branches of science are comparable, we would, from the very beginning, give up the thesis of the unity of science and would regard the process of differentiation as the disintegration of the unity. This atomistic view excludes the formation of the interdisciplinary fields of research and makes, somehow, each discipline similar to a mental microcosmos (Leibniz' monads). Supporters of the unlimited comparability, on the other hand, widen extremely the scope of the comparable mental products and are convinced that mathematics applied to the different branches of science plays the role of a measure of value just like money functions in the different fields of economic life.

If a certain scientific problem is formulated in a too ambitious form the chances of its solution decrease. We think that (Pr 1), being a far-reaching and complicated problem, cannot be given an adequate answer within the scope of this short essay. Therefore, we suggest an answer to a more moderate problem:

(Pr 1′)   When are the hypotheses of a given branch of science comparable?

Problem (Pr 1′) is a real part of the problem (Pr 1), consequently its hypothetical answers are less informative than the solutions of (Pr 1) but are sufficient for making the first step to disentangle the (Pr). We have to see clearly that (Pr) is not about the arbitrary comparison of the hypotheses, on the contrary, it is about their comparison from the point of view of their degree of development.

Before starting to answer the problem (Pr 1′) let us see some examples. Although the hypotheses listed now belong pair-wise to the same field of knowledge the legitimacy of their comparison seems very dubious:

(1a)   Living matter has come to being from dead matter as it is described in Oparin's theory.

(1b)   Man has emerged from animal life as it was thought by Engels.

(2a)   The Hungarian conquest took place in the 890s.

(2b)   The Hungarians had no chance against the Turks in the Battle of Mohács in 1526.

We are sceptical about the comparability of the hypotheses (1a), (1b) and the hypotheses (2a), (2b): we feel that a certain homogeneous substance of knowledge is missing that would guarantee the epistemological affinity of the given hypotheses. In other words: the mentioned hypotheses as possible answers to definite problems are epistemologically almost independent from each other. Belonging to the same branch of science cannot be called a sufficient criterion of comparability. Furthermore, because of the historically incidental character of the borderlines of certain disciplines we do not even recommend it as a necessary criterion.

There is a greater chance to answer the problem (Pr 1′) if the scientific problem-situations arc taken as the basis of comparability of the hypotheses. Each branch of science is a rather heterogenous heap of knowledge; but the scientific problems often going beyond the scope of the given discipline are, in fact, lack of knowledge bounded by on the whole satisfactorily homogenized fields of knowledge. Obviously, the scientific problems are not untouched by the development of knowledge but, according to the moral of the history of science, they show a greater invariance towards change than the hypotheses proposed as their solutions. The problems of the ancient philosophy of nature are less different from the problems of modern physics, chemistry, biology, etc. than as their old and new answers differ from each other. Ancient problems are the forebears of modern problems, the modern ones are, in turn, the succcssors of the ancient problems.

We make a suggestion to introduce the concept of epistemological affinity.

(D 1)   Epistemological affinity holds among two or more hypotheses if, and only if they give an answer to the same problem or problems standing in ancestor-successor relationship.

With the help of the concept of epistemological affinity we can find a rational limit to the comparability of the hypotheses, that is, the pragmatically applicable necessary condition of comparability can be obtained:

(D 2)   Two or more hypotheses are comparable according to their degrees of development if they are in the relation of epistemological affinity.

From (D 1) and (D 2) it follows that:

(D 3)   Two or more hypotheses are comparable according to their degrees of development if they are either rivals or ancestor-successor hypotheses.

According to what has been said above the different theories of light, for example, are comparable. Hypotheses set up about the formation of the Solar system, the emergence of living matter, the origin of mankind, the generation of language, the conquest of the Hungarians, etc. are also comparable.

The view that the scientific problems constitute the basis of the comparability of the hypotheses was put forward by E. Sober (Sober, 1975, 15—33) and I. Levi worked out a theory close to Sober's (Levi, 1967). In Levi's opinion the ultimate basis of the comparability of the hypotheses is that fine structure which characterizes the qualitative manifoldness of the given field of research. It is easy to find, in the background of Levi's theory, the state descriptions and structure descriptions of the Carnapian inductive logic although Levi, contrary to Carnap, does not render a priori numerical value to the predicates representing the qualitative manifoldness of the world (Carnap, 1950). Going back to the origin of Sober's problem-theoretical approach we get to Keynes' principle of the limited variety (Keynes, 1921).

We think that Sober's concept is really suitable for the determination of the classes of the comparable hypotheses since the problems as cognized lacks of knowledge give, in fact, a certain division of the universe in question. For the sake of simplicity we make a difference here only between *solution* problems and *decision* problems. Solution problems block, with the help of one or more categories, (like thing, person, space, time, cause, end, etc.) the unknown; decision problems or value-choosing problems, on the other hand, cover the universe of discourse by setting up the alternative possibilities of solution (hypotheses).

(D 2) as a quasi-necessary condition defines approximately well the scope of comparability; but to say merely that there exist hypotheses corresponding to it is not an adequate answer to (Pr 1'). Since (D 3) describes two alternatives as the condition of comparability it seems reasonable to examine them separately.

If two or more hypotheses answer the same problem they are called *rival hypotheses*. Competition among the hypotheses may be minimal, in this case we have such rhetorical turns that make logically equivalent statements. Competition reaches its maximum when the relation of logical disjunction (antivalence) holds among the hypotheses. So, the rival hypotheses are somehow of the same nature; they are hypothetical knowledge belonging to the same "generation" whose comparability is indubitable. Therefore, the rival character is a sufficient condition of the comparability of the hypotheses:

(D 3.1)   If two or more hypotheses are rivals then they are comparable according to their degrees of development.

The following pairs of hypotheses are good examples of (D 3.1): the Newtonian corpuscular theory of light and Huygens' wave-theory; the conception of Oparin about living organisms and that of Bernal; the interpretation of the relativistic effects by Lorentz—Jánossy and by Einstein. Often more than two hypotheses are competing. It is well known that the theories explaining the formation of the Solar system can be classified into three groups: the turbulence-theories (e.g. Weizsäcker), the captation-theories (e.g. Schmidt) and the catastrophe-theories (e.g. Jeans).

These examples are sufficient to establish the next existential proposition:

(E 1)   According to the evidence of the history of science there exist rival hypotheses.

On the basis of (D 3.1) and (E 1) a conclusion obtains which is a satisfactory answer to the problem (Pr 1'):

(R 1)   According to the evidence of the history of science there exist hypotheses comparable in respect of their degrees of development.

The truth of the answer (R 1) as a presupposition is the minimum condition of the reality (Pr). However, the significance of this answer is not to be overestimated. If exclusively the rival hypotheses are comparable then the answers given to the different scientific problems in the course of the historical development of knowledge, that is, the ancestor-successor hypotheses, are not comparable. In this case, even if we rejected this conception of development formulated by T. Kuhn (Kuhn, 1962), we would be forced to give up the idea of comparison of the hypotheses that have been worked out on the different cognitive levels marked off by the scientific revolutions as cognitive caesuras. Nevertheless, the intuitive conception of the development of science untouched by the speculations of the philosophy of science suggests that science develops rapidly precisely under the stimulation of scientific revolutions. To ignore this circumstance, which is very important from the point of view of our subject matter, would diminish the value of the suggested solution.

Next we shall introduce a concept which gives the condition of the comparability of the ancestor-successor hypotheses. It is impossible, however, within the limits of this study, to outline the entire comparability-theory of A. Schramm, an Austrian logician (Schramm, 1977, 24—44). Here it is sufficient to give the definitions that form the theoretical framework of his theory.

(D 3.1′)   *A* and *B* consistent hypotheses are superficially comparable if, and only if their union is consistent and their product is empty.

(D 3.2)   Hypothesis *A is interpretable* within the scope of hypothesis *B* if, and only if the hypothesis *B* can be supplemented by a set of knowledges *D* to make an hypothesis *B** such that

    (i) *B** is a common extension of both *A* and *B* and each constant of *B** belongs to both *A* and *B*;

    (ii) *D* is the set of propositions valid in *B** such that these propositions are the possible definitions in *B* of all the non-logical constants of *A* not occuring in *B*;

(iii) All the non-logical constants of A not belonging to B occur only in one proposition of D;

(iv) All the propositions valid in B^* can be deduced in B^* from a set of propositions either valid in B or belonging to D.

(D 3.3) Let A and B be consistent and non-trivially comparable hypotheses. In this case, A can be reduced from B if, and only if A is interpretable in B.

(D 3.4) Theory A is *relatively interpretable* within the scope of theory B if, and only if there exists a one-place predicate P not belonging to A such that the A^P version of A relativized to P is interpretable in B.

(D 3.5) Let A and B be incomparable theories. In this case A is reducible from B if, and only if A is relatively interpretable in B.

(D 3.6) Let A and B be such consistent and non-trivially comparable or non-comparable hypotheses which are in ancestor-successor relation. Such hypotheses are comparable according to their degrees of development if, and only if one is reducible from the other.

Hypotheses comparable according to (D 3.3) can be found only in the fields of the abstract-formal sciences (logic and mathematics). Here some propositions of B can be mapped structure-preservingly onto the propositions of A and, using up the definitions in D of the A-constants, propositions of A can be deduced from B. Let us suppose a theory A in which the propositions $S_1, \ldots, S_n$ are valid but the proposition Q is not valid. Then we formulate a theory B such that the valid propositions of A, with the help of their definitions in D, can be translated into the $S_1', \ldots, S_n'$ propositions of B and Q is also valid. Here the $S_1, \ldots, S_n$ are called permanence-principles.

From the point of view of our topic comparability based on relative interpretability has a greater significance. Its necessary and sufficient condition is the P relativizing predicate. This conception has been worked out by R. A. Eberle and G. Vollmer. (See Schramm, 1977, 33—34.) Eberle gives the relativizing predicate between the Newtonian and the Galilean laws of gravitational acceleration. Vollmer has introduced a series of relativizing predicates in a lecture, among others, the relativizing predicate between Kepler's third law of planetary motion and the Newtonian celestial mechanics; between the special relativity theory and classical physics; between the general and the special theories of relativity; between wave optics and geometrical optics; between statistical mechanics and thermodynamics. These relativizing predicates are, in fact, such contrafactual and, at the same time, ideal conditions that in case of their fulfilment the reduced theory is valid.

On the basis of (D 3.6) and the listed examples, the (R 1) as an answer to (Pr 1′) can be supplemented as follows:

(R 1′)   According to the evidence of the history of science there exist such pairs of hypotheses of which the ancestor-hypothesis can be reduced from the successor-hypothesis and so they are comparable in respect of their degrees of development.

The (R 1′) is a more pretentious supposition than (R 1) and, at the same time, it offers a possibility to make (D 3) more exact:

(D 3′)   Two or more hypotheses are comparable according to their degrees of development if, and only if they are either rivals or, as ancestor-hypotheses, are reducible from certain successor-hypotheses.

3 THE FACTORS OF THE DEGREE OF DEVELOPMENT

Now we start to solve the problem (Pr 2). In the following, the hypotheses considered will be supposed to be comparable. If two hypotheses are comparable, then, within the scope of the present conception, we do not raise the question whether one is more developed than the other or that they are equally developed. Obviously, we are not saying that it is completely impossible to work out a theory that would make the hypotheses unlimitedly comparable; this, however, is not the goal of the present discussion.

126

If we want to set up the hierarchy of two or more hypotheses from the point of view of their degrees of development then we cannot avoid to answer the question as to what are the *most important characteristics of development* since one hypothesis should surmount the other in respect of these characteristics to be qualified as the more developed hypothesis.

It would be best to answer the emerged problem first on the philosophical level. We have to do so, because "development" is a philosophical category and, moreover, the degree of development of the hypotheses is an epistemological question.

Here we are not going to analyze the numerous philosophical conceptions concerning development. Even the task of listing the current views of the Marxist scholars would go beyond the limits of this short essay. Instead, we shall concentrate on the elaboration of our conception about the development of scientific knowledge and the evaluation of our enterprise is left to the reader.

The starting point is that each scientist-generation finds a rather heterogeneous complex of scientific knowledge at hand that must be acquired selectively with the help of the previous generation in order to rise to the level of *general culture.* This level of knowledge constitutes the background of *professional culture* which has several stages from professional skill to the level of knowledge demanded by scientific research. The latter is equal with the prevailing world level of human knowledge or at least it does not deviate from it significantly. This level can be reached only by the best thinkers of mankind and they really know where the current limits of human knowledge are and what those knowledge-needs are whose satisfaction is indipensable, or at least desired from the point of view of human progress. Being aware of these knowledge-needs the scientists can *realize and conceptualize the scientific problems.*

Scientific problems as the *realized cognitive needs* of mankind create a special *tension-situation* that stimulates the human mind for a kind of mental seeking. This tension holds between the so-called reality-level of cognition and its need or the desire-level outlined on the reality-level but existing only in the form of possibility.

As electric tension shows the lack of balance in the distribution of the charges, or as social tension demonstrates the lack of balance among the political powers of society, similarly, scientific problems represent the tension between the status quo of knowledge and the sphere of possibility grounded by it.

We think that the interpretation of the problems as cognitive tension-situations promotes a better understanding of the nature of *dialectical contradiction*, at least in the field of the development of science. Now we sketch our conception only in a qualitative form in order to formulate our suggestions in the first draft.

First, the concept of cognitive tension-situation must be clarified. To do so, we have to reveal those factors which determine the tension-situation. Our starting point is that science is a special dynamic system having a well-defined inner structure conforming through its functions more or less adequately to those "surroundings" which constitute the field of play of its formation and application. The first is called the *inner value* or perfection of scientific knowledge; the second is called its *empirical adequacy*. Inner perfection is basically a structural characteristic, empirical adequacy is of a functional nature. Perhaps there is no need of arguing to support the statement that the two factors mentioned influence the *degree of development* of the given system of knowledge (the hypothesis) which is the third significant parameter from the point of view of our present discussions.

In the elaboration of the development-theory of scientific knowledge a fourth parameter plays an important role, too, which is called the *sharpness of the problem-situation* or the degree of its problematic character. In cognition the earlier knowledge as reality and the new knowledge to be acquired as a mere possibility constitute a dialectical unity. This unity, which at the same time contains a certain aspect of exclusion, manifests itself in the sharpness of the problem-situation. This parameter anticipates somehow the work of cognition of the epistemical subject that must be fulfilled in order to diminish the gap between the increased cognitive needs and the given reality-level of knowledge.

There can be imagined such a cognitive situation, when, for example, a greater degree of empirical adequacy is required and we are satisfied with the previous level of the inner value, or perhaps, with a lower level, too. The opposite situation may occur as well: we raise a higher demand concerning the inner value of the new problem-solution and admit the stagnation or even a certain decrease of its empirical adequacy.

So the degree of development of the new hypothesis might be less than that of the old one but it cannot fall under a certain critical value. This seems obvious, since the new hypothesis must meet a higher level of need, that is, it must give an answer to a more difficult problem compared with the task of the old hypothesis. A threshold-value of the degree of development can be given that functions as the value-norm of acceptability on the current level of knowledge. Therefore, scientific hypotheses cannot be compared in respect of their degrees of development independently from the cognitive situations.

Let us take a hypothesis h_1 in a certain cognitive situation n. If it satisfies the desirable level of empirical adequacy and inner perfection better than all the other hypotheses that count, then it is accepted as an answer to the investigated problem. In these cases the scientific public opinion has a great confidence in the accepted hypothesis and becomes resistent towards the falsifying empirical facts that emerge and is tolerant with the theoretical difficulties of the conception. This resistent phase often lasts long but it cannot be eternal. After some. time the dissatisfaction of the scientific public opinion with the given hypothesis reaches a level where it withdraws its intellectual confidence and shows a critical attitude. So a new cognitive cycle begins that goes together with the reformulation of the problem and with the seeking of new answer-possibilities.

In the new cognitive situation $n+1$ we are searching for such a solution alternative h_2 which is a better answer to the investigated problem than h_1 was before it has lost its status as an answer. So we demand that the degree of the problematic character of the hypothesis h_2 in the cognitive situation $n+1$ should be less than that of h_1 in the cognitive situation n, or at least it should not increase.

We know very well that the conception outlined here presupposes an idealized development of science in which the epistemic values of the scientific development fully prevail. In the scientific autobiographies (epistemological credo) of the best scientists the view is often unambiguously formulated that to be a scientist means to enrich mankind with new, valid and systematic knowledge. The scientist undertakes the task of setting up new solution-alternatives. But he does not trust the new theoretical constructions and therefore seriously supervises them. If the hypotheses are not isolated but are systematically related to one another this revision and the conveyance of knowledge become easier.

The present logical approach can always be questioned by confronting it with historico-empirical facts. However, we have to be very cautious when we decide to give up. It might happen, for example, that because of some falsifying facts we have to abandon our claims to explain scientifically and predict facts of greater importance, and our earlier systematical knowledge splits up into a loose conglomerate of isolated factual statements and low-degree hypotheses.

Ideal models always contain something unattainable; but their lack deprives knowledge from those values with which the status quo of knowledge must be faced. These long-term goal-values of scientific research are often called methodological ideals, paradigms or research programs.

To sum up our discussion: the degree of development of the scientific hypotheses is uniquely determined, in the solution-phase, by empirical adequacy and inner perfection.

Our claims concerning empirical adequacy and inner perfection may increase in a new cognitive situation where the earlier hypothesis does not meet the requirements any more. However, we must be tolerant at the renouncement of the old hypothesis while a more adequate problem-solution is not found. Here the new problem has already been formulated since the difference between the old knowledge and the new cognitive need became evident.

These considerations are of a qualitative nature and their task is no more than to motivate and prepare the quantitative mode of discussion in the next chapter.

Summing up the conclusions of this chapter, we can answer the problem (Pr 2) as follows:

(R 2)   There are four parameters playing an important role in the structure of the theory of development of science: empirical adequacy, inner perfection, the actual level of development, and the degree of sharpness of the problem-situation.

4 RELATION AMONG THE VARIABLES DETERMINING THE DEGREE OF DEVELOPMENT

Before suggesting an adequate mathematical formula for the establishing of a hierarchy of the hypotheses according to their degrees of development, it seems reasonable to formulate some intuitively justifiable requirements whose fulfilment is required from the given formula.

a) The formula must be such that its degree of development is the greater as the empirical adequacy of the hypothesis under discussion is greater, supposing that inner value is constant.

b) The formula must have a phase within which the increase of the inner value decreases the degree of development of the hypothesis and it must also have a phase in which the inner value increases the degree of development, supposing that empirical adequacy is constant.

c) The required formula must have a zone within the mentioned intervals such that the value of the degree of development rapidly increases in it from a low level to a relatively high one.

d) Let the jump in the mentioned zone be greater the greater the empirical adequacy and the inner perfection.

It is not difficult to see that in the requirements given by a)—d) the wanted curve must be such as to describe the relatively slow quantitative phases of development [a) and b)] and, at the same time, it must record the relatively fast, new quality-creating phase of development [c) and d)]. Therefore, our goal is to connect the problem-situation as a cognitive contradiction with the dialectics of quality and quantity.

In order to limit our discussions it is reasonable to introduce another restriction based on the presupposition that the development of science is, with a relatively good approximation, the sequence of cycles following upon one another in time. These cycles, roughly speaking, are the repeated occurrences of a certain scientific problem in historically sharply different cognitive situations, where these occurrences of the problem are solved by paradigms of higher and higher levels comparable with each other. According to what has been said above a further restriction seems necessary to be introduced:

e) The formula we are searching for is required to describe only one cycle of development abstracting from the long run historical succession of the cycles of development.

The conceptual framework of our discussion is constituted by the so-called *problem-theoretical model* of scientific knowledge. Consequently, we have to find a function which not only fulfils the given a)—e) conditions, but also gives a possibility to measure the sharpness of the problem as a cognitive tension-situation. It can be seen easily that the sought mathematical formula can be only a potential function holding among the next variables: the sharpness of the problem-situation, the degree of development of the actual knowledge, the empirical adequacy of knowledge fixed by the methodological ideal and its inner perfection.

Let p be the sharpness of the problem-situation, e the degree of development of the knowledge at hand, u the inner perfection and v the desired degree of empirical adequacy. We think that the potential function

$$p = 0.5e^4 - 0.5ue^2 - ve \tag{1}$$

is adequate to meet the mentioned requirements. This function is in fact the peak catastrophe function of R. Thom (Thom, 1972). Only the $[-1, 1]$ interval of the function (1) is interesting for us, because -1 is taken as the minimum and 1 as the maximum sharpness-value of the problems. In other words, the function (1) describes one cycle of the problem-solution in the $[-1, 1]$ interval. In this case the values of e and v also fall within the $[-1, 1]$ interval and the values of u are in the $[0, 1]$ interval. However, further considerations are necessary to a satisfactory interpretation of the minimum and the maximum values of the mentioned parameters.

Sciences, metaphorically speaking, strive to draw in to an e level where tension is minimal. These minimum-places of the potential function are called attracters. Science as an open system approaches to one of these asymptotically stable places with the growth of time. These equilibrium places can be interpreted as the possible solutions of the scientific conflicts (problems) represented by the potential function. The group of these possible solutions is the M_p set of the minimum-places of the gradient of the function (1) and it can be described by the equality:

$$p' = 2e^3 - ue - v = 0 \qquad (2)$$

The equation (2) is represented by the next figures:

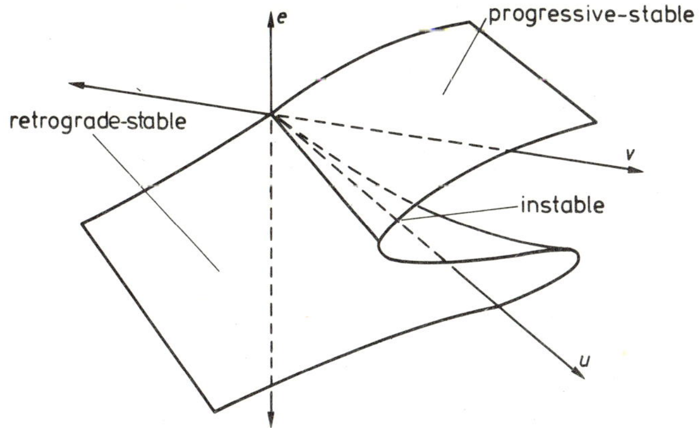

Figure 12

The projection of the threefold part of the surface M_p on the parameter-plane (u, v) is the shaded plane-part K_p (figure 12) which can be described by the equality:

$$27v^2 - 2u^3 = 0 \qquad (3)$$

The K_p set of points divides the parameter-plane (u, v) into two parts: above the unshaded outer parts and below them the surface M_p is onefold. In these places the system has only one attracter. The inner shaded plane-part is the project of the threefold part of the surface M_p, in the points of the upper and lower levels of $K_p p$ has its minimum, in the points of the middle level p has its maximum. So the system has two attracters in the points of the plane-part "within" the K_p set that are equally strong at about the middle of the shaded plane. The system is instable in the points of the plane-part K_p. If we proceed in K_p "from outwards to inwards" then, except the peak, a new and in the beginning

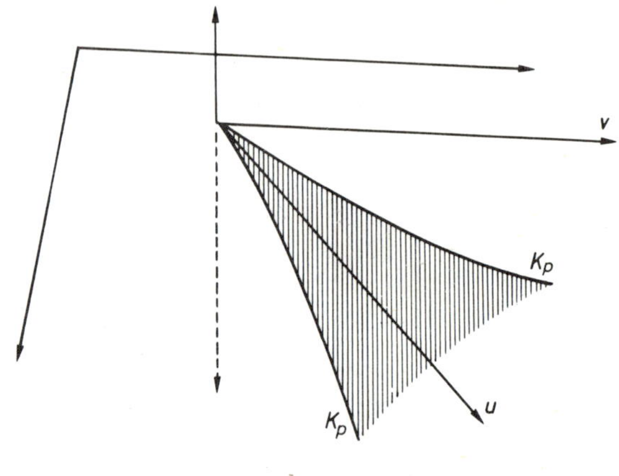

Figure 13

weak attractor is generated which is becoming stronger as we proceed in K_p. And proceeding "outwards", the new attracter becomes dominant but the system jumps into the state marked by the new attracter only when we cross the borderline of the plane-part K_p.

Consequently, the variable v used to be called normal parameter and the variable u called dividing parameter. The latter name is really appropriate since the greater the value of u, the greater the distance between the two attracters.

Applying what has been said above to the process of the development of science we can make the following remarks:

a) If the empirical adequacy of a hypothesis with $u = 0$ minimal inner perfection increases then its degree of development undergoes a gradual qualitative change. Within this mode of development the retrograde-stable phase and the progressive-stable phase do not separate from each other and the instable state shrinks point-like. This mode of development characterizes, with a good approximation, sciences based on loosely connected empirical regularities. These sciences well approximate the *cumulative model of development* unfavourably mentioned by Kuhn.

b) If the inner value u of the hypothesis under discussion is greater than the minimum then its development is rapid which happens when the value of the empirical adequacy v falls within the

$$\left[-\sqrt{\frac{2u^3}{27}}, \sqrt{\frac{2u^3}{27}} \right] \text{ interval.} \tag{4}$$

Supposing that $u=1$, we get for v the value $\pm 0,2722$. This result can be interpreted as follows: if the inner value of a theory is maximal then its empirical adequacy must reach a definite positive value in order that it could give a progressively stable solution of the problem; and it must have an opposite negative value in order that it could give a retrograde-stable solution of the problem. This means, that a well-structured hypothesis can be called a theory only above a certain limit of value of its empirical adequacy and it may be refused as a completely unsuccessful hypothesis only in case of a relatively small degree of adequacy. The given formula differentiates well, at a high inner value, between empirically adequate and empirically non-adequate solutions. In the theories that are innerly less perfect this differentiation is smaller. This result is in accordance with the intuitive endeavour to promote the structural perfection of science.

c) Changes taking place either in the progressive-stable phases or in the retrograde-stable phases are of a quantitative nature. Changes in the retrograde phase are not sufficient to bring the hypothesis in accordance with the changed cognitive needs. Changes within the progressive phase increase the harmonization of the hypothesis with the ideal of science. The instable phase of development is always the symptom of a *qualitative* change whose rapidity is the greater the greater the inner value of the hypothesis is and as the empirical adequacy approximates the value 0.

d) If h_1 and h_2 are hypotheses answering the same problem (Pr) and the degree of development of one hypothesis falls into the retrograde-stable phase and the degree of development of the other hypothesis falls into the progressive-stable phase, and, moreover, the hypotheses are comparable in the previously defined sense, then the degrees of development of these hypotheses are qualitatively different and the more developed hypothesis is the *dialectical negation* of the less developed one.

The great explanatory power of the conception outlined here manifests itself in the fact that it gives a possibility for a modern interpretation of the *dialectical negation*. As far as I know, no one has succeeded, up to now, in giving such a formally also correct explication of this dialectical category so mysterious for many people that would surpass the present conception.

We think that this conception can be accepted as a global answer to the problem (Pr 3) whose essence is summed up as follows:

(R 3)   For the description of the development of science the fourth order potential function of R. Thom representing the peak catastrophe is well applicable. This function describes, in a handy form, the relation among the next parameters of development: the sharpness of the scientific problem, the degree of development of the answer given to the problem, the desirable degree of empirical adequacy and inner perfection required from the answer. The possible solutions (hypotheses) are pointed out by the minimum places of the given function. The model of development outlined here gives a possibility for a modern interpretation of the dialectical contradiction, the dialectical connection between quality and quantity, and the dialectical negation.

It is easy to recognize, however, the basic phenomenological character of this model of development and that it does not say too much about the functional mechanism that governs the scientific development. In order to remove this deficiency we have to eliminate the black-box character of the theory. We do not, however, undertake here the solution of this problem.

136

5 CHOICE BETWEEN THE SCIENTIFIC HYPOTHESES

Before seeking an answer to (Pr 4) we make a suggestion for the determination of the numerical values of the parameters u and v that are very important from the aspect of the degrees of development of the hypotheses. It must be remarked here that the forthcoming suggestions are independent from the previously outlined theory. It can be imagined very well that the suggested operational definitions can be replaced by more adequate ones, but the theory can be applied successfully only with appropriate measuring instructions.

Let us look first at the parameter v! Our starting point is that empirical adequacy is in fact the relative truth value for the measuring of which we can apply the formula suggested by Bunge (Bunge, 1967, vol. 2. 102—103):

$$V = 0.5 + \frac{C + (R - C)\varepsilon}{2N} - \delta \qquad (5)$$

where N is the number of all the experientially testifiable consequences of the hypothesis ($N \gg 1$), C is the number of the confirmed cases and R is the number of the refuted cases ($C + R = N$), ε is the measure of the empirical errors and δ is the number of the theoretical (logical and/or mathematical derivational) errors. It can be seen easily that V takes values in the $[0, 1]$ interval. The values of the parameter v fall in the $[-1, 1]$ interval. In order that V could be adequate for the measuring of the empirical adequacy, the equality (5) must be transformed according to $v = 2V - 1$

$$v = \frac{C + (R - C)\varepsilon}{N} - 2\delta \qquad (6)$$

(6) expresses the so-called relative covering-value of the discussed hypothesis. The testifiable facts N deduced from the hypothesis are taken here as having equal weights. Obviously, these facts are not individuals but fields of facts, more or less well separable from each other, that are covered by the given hypothesis. However, Bunge's approach seems unsatisfactory in two respects: 1. We think that the

theoretical error can be ruled out supposing that logical error and/or mathematical miscalculation is not committed. Theoretical errors do not necessarily accompany the empirical cognitive activity. 2. At the same time the formula (6) does not take into consideration the factor which, after László Ervin, can be called resistance factor. (E. László, 1972, 379—395). It expresses the resistance of the scientific workers against the innovations or, in other words, their orthodox attitude. If the scientific public opinion is resistant towards those deviations which refute the theory then it is practically satisfied with the status quo of science.

Next we shall make a suggestion on the introduction of a new operational definition. It is supposed that empirical adequacy is the greater the number of the confirming cases is the greater and the resistance-factor of the scientific public opinion is the greater; and empirical adequacy is the smaller the number of the refuting cases is the greater and the empirical error and the number of the experientally testifiable cases are the greater. The next formula satisfies these conditions:

$$v = \frac{(C-R)(1-\varepsilon)\sigma}{N}. \tag{7}$$

It is supposed that the value of the empirical resistance-factor σ changes in the [0, 1] interval and it has a maximum value in the period of acceptance of the hypothesis. Its value decreases rapidly only when the refuting cases could not be eliminated despite of all theoretical efforts and a new successor hypothesis has emerged whose adequacy is greater than that of the earlier hypothesis.

Let us examine next the possibility of metrization of the parameter u. I think we have to face here a more difficult problem than before.

J. T. Davies gives the next formula for the measuring of the inner value (perfection) of the hypotheses (Davies, 1973, 162—164):

$$u = 0.005\, G^2 ST \tag{8}$$

where G is the generality, S is the simplicity, and T is the exactness of the predictions of the hypothesis. The value of each mentioned parameter may change within the [0, 5] interval. These values are comparatively estimated. The 0.005 is a factor of proportionality which limits the values of u to the $[0, 1]$ interval. However, it must be also required that the weak unequality

$$G^2 ST \leqq 200$$

should be satisfied.

We make a suggestion for a smaller modification of the formula (8): we introduce the ρ theoretical resistance-factor which expresses our tolerance towards the weaknesses of the theory. Obviously, the inner value u is the greater our tolerance towards the theoretical errors is the greater. From this it follows

$$u = 0.05 \, G^2 ST\rho . \tag{9}$$

Now we shall describe our comparative value-tabulation made after Davies. We remark that the parameters C and R do not, in the forthcomings, refer to the actual number of the confirming and the refuting cases but merely to their values on points when the next connection obtains: $C + R = N = 5$.

C confirmation

"$\pi = 3.1415926\ldots$"	5	"All swans are white"	2
"The Earth is round"	5	Relativity theory	4
Laws of thermodynamics	4.5	Weather-forecast	2.5
"The Earth has a perfectly round shape"	1	"This swan is white"	3
Atomic theory	5	Astronomical theories	2
Newtonian mechanics	4	"There is life on other planets"	0.5

R refutation

"$\pi = 3$"	5	The atomic theory	0
"The Earth is perfectly round-shaped"	4	The theory of thermodynamics	0
"All swans are white"	2	The relativity theory	0
Weather-forecast	1.5	"This swan is white"	0
The Newtonian mechanics	1	"The Earth is round"	0

| Astrological theories | 1 | "$\pi = 3.1415926\ldots$" | 0 |
| The law of radioactive half-period | 0.5 | | |

G integrating power

Unified theories of space	5	"All swans are white"	1.5
Relativity theory	4	"The Earth is round"	1.5
Newtonian mechanics	3	Astronomical theories	2
Atomic theory	3	"This swan is white"	0
The laws of thermodynamics	3	"$\pi = 3.0$" (exactly)	0
Weather-forecast	2	"$\pi = 3.1415926\ldots$"	0

S simplicity

"$\pi = 3$"	5	The laws of thermodynamics	3
"The Earth is perfectly round-shaped"	5	The relativity theory	2
"$\pi = 3.1415926\ldots$"	4	"The Earth is round"	2
The atomic theory	4	Weather-forecasts	1
The Newtonian mechanics	4	Computer-curve of data	0
"This swan is white"	4		
"All swans are white"	4		

T exactness

The computer curve of the data	5	The weather-forecasts	2
"All swans are white"	4	"The Earth is round"	1.5
The Newtonian mechanics	4	The astrological theories	1
The relativity theory	4	$= 3.1415926\ldots$	0
The atomic theory	4	"This swan is white"	0
The laws of thermodynamics	3	Some psychoanalitical theories	0
"The Earth is perfectly round-shaped"	2	Simple mathematical and logical connections	0

Formulas (7) and (9) make it possible to revise our theoretical considerations examining some concrete cases.

Let us see the Newtonian mechanics and the relativity theory! Newtonian mechanics has been regarded as indubitable before the formulation of the relativity theory, although some facts contradicting to it were well-known. In other words: the scientific public opinion confidently believed that the difficulties emerged can be solved, sooner or later, on the basis of the theory. Consequently the values of the resistance-factors σ and ρ are high in the given cognitive situation. Let us suppose that their values are equally 0.95. And let the measuring error ε be 0.05. Substituting the proper values into the formula (7) we get the following result for the Newtonian mechanics in the cognitive situation n:

$$v(m)_n = \frac{(4-1)0.95 \cdot 0.95}{5} = 0.48 \, .$$

Let us, then, imagine ourselves in the $n+1$ cognitive situation which came about with the formation of the relativity theory. The empirical resistance-factor of the scientific public opinion decreases nearly to the minimum. The theoretical resistance-factor does not change since the Newtonian mechanics as a theoretical production invariably preserves its value. Let $\varepsilon = 0.05$ in the $n+1$ cognitive situation:

$$v(m)_{n+1} = \frac{(4-1)0.05 \cdot 0.95}{5} = 0.0285 \, .$$

Let us compute the empirical adequacy of the relativity theory in the $n+1$ cognitive situation:

$$v(r)_{n+1} = \frac{4 \cdot 0.95 \cdot 0.95}{5} = 0.72 \, .$$

The inner values of the examined theories are computed in a similar way in the cognitive situations n and $n+1$. It is supposed that the theoretical resistance-factor has a nearly maximal value:

$$u(m)_n = u(m)_{n+1} = 0.005 \cdot 3^2 \cdot 4 \cdot 4 = 0.72$$

$$u(r)_{n+1} = 0.005 \cdot 4^2 \cdot 2 \cdot 4 = 0.64 \, .$$

Substituting the obtained values into (2) we can compute the degree of development of the Newtonian mechanics and the relativity theory in the given cognitive situations:

$$e(m)_n^3 - 0.5 \cdot 0.72 \, e(m)_n - 0.5 \cdot 0.48 = 0$$

$$e(m)_{n+1}^3 - 0.5 \cdot 0.72 \, e(m)_{n+1} - 0.5 \cdot 0.0285 = 0$$

$$e(r)_{n+1}^3 - 0.5 \cdot 0.64 \, e(r)_{n+1} - 0.5 \cdot 0.72 = 0 \, .$$

By the solutions of the equations we get here:

$$e(m)_n = 0.81$$

$$e(m)_{n+1} = \begin{cases} -0.62 \\ -0.04 \\ -0.58 \end{cases}$$

$$e(r)_{n+1} = 0.78 \, .$$

Comparing the degrees of development we obtain the following conclusions:

1. With the fundamental change of the empirical resistance-level of the public opinion the Newtonian theory got from the progressive-stable phase of development to the instable phase.

2. The degree of development of the relativity theory is smaller than that of the Newtonian theory was in the progressive-stable phase but it also falls in the progressive-stable phase of development.

Interesting results obtain when the degrees of the problematic character of the examined theories are computed with the help of (1):

$$p(m)_n = 0.5 \cdot 0.81^4 - 0.5 \cdot 0.72 \cdot 0.81^2 - 0.48 \cdot 0.81 = -0.58$$

$$p(m)_{n+1} = 0.5 \cdot 0.619^4 - 0.5 \cdot 0.72 \cdot 0.619^2 - 0.0285 \cdot 0.619 = -0.082$$

$$p(r)_{n+1} = 0.5 \cdot 0.78^4 - 0.5 \cdot 0.64 \cdot 0.78^2 - 0.72 \cdot 0.78 = -0.76 \, .$$

Comparing the results our conclusions are as follows:

3. With the decrease of the resistance-factor the degree of the problematic character of the Newtonian mechanics has considerably increased.

4. The degree of the problematic character of the relativity theory in the cognitive situation $n+1$ is smaller than that of the Newtonian mechanics was in the cognitive situation n.

Let us introduce the formula $1-p$ for the denotation of the *degree of solution of the problem*. Then the result 4. sounds like this:

4′. Relativity theory provides a more satisfactory problem-solution in the cognitive situation $n+1$ than what was afforded by the Newtonian mechanics in the cognitive situation n.

Similar computations can be made at the comparison of other theories. Without overestimating the exactness of our results we may say that they show a satisfactory accordance with our intuitive expectations.

The results here obtained make it possible to answer the (Pr 4).

(R 4)   (i)   Only such a hypothesis can be accepted as an adequate problem-solution on the given level of knowledge whose degree of development falls in the progressive-stable phase, that is $e > e_0$
where e_0 is the root of the equation

$$e_0^3 - 0.5\, ue_0 - 0.5\,\frac{2u^3}{27} = 0 \quad (e_0 \approx 0.8164 \,|\sqrt{u}|\,).$$

(ii)   Only such hypothesis can be accepted as an adequate problem-solution on the given level of knowledge whose degree of the problematic character is smaller than that of the rival hypothesis or that of the earlier accepted ancestor-hypothesis comparable with it.

(iii)   If we can not choose among the hypotheses that count by applying (i) and (ii) then we choose the hypothesis with the greatest empirical adequacy.

(iv)   If no difference can be pointed out in the hypotheses according to (i)—(iii) then we choose the hypothesis with the greatest inner value.

(v)   If the criteria (i)—(iv) do not work then we choose at will among the proposed conceptions.

(vi) If none of the given hypotheses are acceptable but there is one whose degree of development is greater than 0, then our attitude towards it is *positively non-committal* with the hope that its degree of development can be raised to the desired level by applying appropriate subsidiary hypotheses.

(vii) If more hypotheses meet the condition (vi) then we decide according to the rules (ii)—(v).

(viii) If among the hypotheses under discussion there is not any that could meet the rules (i)—(vii) but there is one hypothesis among them whose degree of development falls in the retrograde-instable phase, that is, $0 > e > -e_0$, then our attitude towards it is *negatively non-committal,* that is, for lack of something better we do not reject it provisionally.

(ix) If there are more hypotheses that meet the condition (viii), then we differentiate them by properly applying the rules (ii)—(v).

(x) In each cognitive situation *only one hypothesis can be accepted.* If there is no such hypothesis then *our attitude can be positively non-committal at most towards one hypothesis.* If the two previous alternatives are ruled out then *our attitude can be negatively non-committal at most towards one hypothesis.* The fourth alternative is that all the hypothesis that have been reckoned with are rejected as retrograde-stable solutions.

6 SUMMARY

In this appendix we have sought an answer to an important problem of the theory of science. It would be da ing to say that our suggestions pass the test of scientific criticism in every respect. Here a series of maturing conceptions have been presented to the reader and we hope that not all of them prove to be theoretical deadlocks. Surely, the greatest doubts are raised by the computing methods. However, these considerations do not constitute an organic part of the theory outlined here and can always be replaced by more appropriate methods.

144

BIBLIOGRAPHY

ALLPORT, G. W. (1954) The Nature of Prejudice. Reading, Mass.: Addison-Wesley.

APOSTEL, L. (1964) 'Rhétorique, psycho-sociologie et logique.' In: La théorie de l'argumentation. Centre National Belge des Recherches de Logique. Louvain-Paris: Ed. Nauwelaerts, 263—314.

BUNGE, M. (1967) Scientific Research I—II. Berlin, Heidelberg, New York: Springer Verlag.

BUNGE, M. (1974) Treatise on Basic Philosophy, I—II. Dordrecht: Reidel Publishing Co.

CARNAP, R. (1950) Logical Foundations of Probability. Chicago: The University of Chicago Press.

CARNAP, R.—BAR-HILLEL, Y. (1964) 'An Outline of the Theory of Semantic Information', In: Language and Information (Ed. by Y. Bar-Hillel), Reading, Mass.: Addison-Wesley.

CHOMSKY, N. (1967) Recent Contributions to the Theory of Innate Ideas. Synthese, *17*, 2—11.

DAVIES, J. T. (1973) The Scientific Approach. London—New York: Academic Press.

FESTINGER, L. (1957) Theory of Cognitive Dissonance. Eveston Mionis: Row, Peterson.

HAMBLIN, C. L. (1971) 'Mathematical Models of Dialogue', Theoria, *2*, 130—155.

HARRAII, D. (1963) Communication: A Logical Model. Cambridge, Mass.: M.I.T. Press.

HILPINEN, R. (1968) Rules of Acceptance and Inductive Logic. Amsterdam: North-Holland Publishing Co.

HINTIKKA, J. (1968) 'The Varieties of Information and Scientific Explanation' In: Logic, Methodology, and Philosophy of Science III, (Ed. by B. von Rootsellar and J. F. Staal) Amsterdam: North-Holland Publishing Co. 311—331.

HINTIKKA, J.—PIETARINEN, J. (1966) 'Semantic Information and Inductive Logic' In: Aspects of Inductive Logic. (Ed. by J. Hintikka and P. Suppes) Amsterdam: North-Holland Publishing Co., 96—112.

HUGHES, G. E.—CRESSWELL, M. J. (1968) An Introduction to Modal Logic. London: Methuen.

KEYNES, J. M. (1921) A Treatise on Probability. London: MacMillan.

KONGER, A. N. (1960) Modern Debate. Its Logic and Strategy. New York, Toronto, London: McGraw-Hill.

KUHN, T. S. (1962) The Structure of Scientific Revolution. Chicago: The University of Chicago Press.

Lakatos, I. (1971) 'History of Science and Its Rational Reconstruction' In: Boston Studies in the Philosophy of Science. Vol. VIII. Dordrecht: Reidel Publishing Co.

László, E. (1972) A General System Model of the Evolution of Science. Scientia. Annus LXVI. Vol. 107. V—VI. 379—395.

László, Gy. (1970) A "kettős honfoglalás"-ról. Archeológiai Értesítő, 1970/2. 161—190.

Levi, I. (1967) Gambling with Truth. New York: Alfred A. Knopf.

Lewin, K. (1935) A Dynamic Theory of Personality. New York, London: McGraw-Hill. 1—42.

Lukács, G. (1963) Werke, 11., 12.: Ästhetik I. Die Eigenart des Ästhetischen. 1—2 Halbband. Neuwied: Luchterhand.

Mannheim, K. (1929) Ideologie und Utopie. Frankfurt am Main: Schulte-Blumke.

Maruyama, M. m (1960) 'Communicational Epistemology.' British Journal for the Philosophy of Science. 11: 319—327; 12: 52—62; 12: 117—131.

Maruyama, M. m (1974) Paradigmatology and Its Application to Cross-disciplinary, Cross-professional and Cross-cultural Communication. Dialectica, Vol. 28. 3—4.

Merton, R. K. (1972) The Institutional Imperatives of Science. Sociology of Science.

Nowak, L. (1976) 'On Some Interpretations of the Marxist Methodology'. Zeitschrift für allgemeine Wissenschaftstheorie. Band VII, Heft 1, 141—183.

Pataki, F. (1977) Társadalomlélektan és társadalmi valóság. Budapest: Kossuth.

Perelman, C.—Olbrechts-Tyteca, L. (1958) La nouvelle rhétorique. Traité de l'argumentation I—II. Paris: Presses Universitaires de France.

Perelman, C. (1968) Éléments d'une théorie de l'argumentation. Bruxelles: Presses Universitaires de Bruxelles.

Pietarinen, J. (1970) 'Quantitative Tools for Evaluating Scientific Systematisations'. In: Information and Inference. (Ed. by J. Hintikka and P. Suppes) Dordrecht: Reidel Publishing Co. 123—147.

Polya, G. (1954) Mathematics and Plausible Reasoning, I—II. Princeton, New Jersey: Princeton University Press.

Planck, M. (1948) Wissenschaftliche Selbstbiographie. Leipzig: Verlag J. Ambrosius Barth.

Popper, K. (1959) The Logic of Scientific Discovery. London: Hutchinson.

Popper, K. (1972) Objective Knowledge. Oxford: Clarendon Press.

Putnam, H. (1967) The 'Innateness Hypothesis' and Explanatory Models in Linguistics. Synthese, *17*, 12—22.

Rescher, N. (1969) Many-valued Logic. New York, etc: McGraw-Hill.

Rescher, N. (1970) Scientific Explanation. New York, London: The Free Press.

Rescher, N. (1973) The Coherence Theory of Truth. Oxford: Clarendon Press.

Rosenboom, W. W. (1971) 'New Dimensions of Confirmation Theory. In: Philosophy of Science Association 1970 Meeting. (Ed. by R. C. Buck and R. S. Cohen) Dordrecht: Reidel Publishing Co.

Russell, B. (1948) Human Knowledge. Its Scope and Limits. London: Simon and Schuster.

SCHRAMM, A. (1977) Wahrheitsähnlichkeit und Reduktion. In: Probleme des Erkenntnis-
fortsschritts in den Wissenschaften. (Ed. by K. Freisitzer and R. Haller). VWÖ Wien,
24—44.

SIMON, A. A. (1968) 'On Judging the Plausibility of Theories.' In: Logic, Methodology and
the Philosophy of Science III. (Ed. by B. von Rootsellar and J. F. Staal) Amsterdam:
North-Holland Publishing Co. 440—459.

SOBER, E. (1975) Simplicity. Oxford: Clarendon Press, 15—33.

THOM, R. (1972) Advanced Book Program Reading. Mass.: W. A. Benjamin Inc.

VON WRIGHT, G. H. (1957a) The Logical Problem of Induction. Second revised edition.
Oxford: Barnes and Noble.

VON WRIGHT, G. H. (1957b) 'A New System of Modal Logic.' In: Logical Studies. London:
Humanities.

VON WRIGHT, G. H. (1970) 'A Note on Confirmation Theory and on the Concept of
Evidence. Scientia, Vol. CV. N. DCCI—DCCII. ser. VII. IX—X.